# EXPLORATION OF CORTICAL FUNCTION

# EXPLORATION OF CORTICAL FUNCTION

## Imaging and Modeling Cortical Population Coding Strategies

by

MARTIN STETTER
*Siemens AG,*
*Munich, Germany*

KLUWER ACADEMIC PUBLISHERS
DORDRECHT / BOSTON / LONDON

A C.I.P. Catalogue record for this book is available from the Library of Congress.

ISBN 1-4020-0435-4 (HB)

---

Published by Kluwer Academic Publishers,
P.O. Box 17, 3300 AA Dordrecht, The Netherlands.

Sold and distributed in North, Central and South America
by Kluwer Academic Publishers,
101 Philip Drive, Norwell, MA 02061, U.S.A.

In all other countries, sold and distributed
by Kluwer Academic Publishers,
P.O. Box 322, 3300 AH Dordrecht, The Netherlands.

*Printed on acid-free paper*



Printed in the Netherlands.

*For Beate*

# Contents

# Acknowledgments

I wish to thank many of my collegues, who worked together with me in developing the ideas and solutions provided in this book, but also in my efforts to merely recognize some of the real world problems to be solved on our way towards a better understanding of cortical function. Many of these problems have been attacked together with Klaus Obermayer and his coworkers. I wish to thank Klaus for his continuous efforts in establishing and maintaining communication and collaboration between theoreticians and biologists and for sharing his vision about how to setup successful interdisciplinary work. Special thanks to the students and collegues I had the pleasure to work with, in particular to Hauke Bartsch, Ute Bauer, Anca Dima, Heiner Greve, Thomas Knop, Thomas Otto, André Paus, Ingo Schießl, Holger Schöner, Michael Scholz and Roland Vollgraf.

I am deeply indepted to Jenny Lund and John Mayhew for introducing me into many important aspects of their own fields and finding an benevolent and patient but at the same time efficient way to do so. They never lost their confidence in our common work and together with their groups showed an untiring effort to make it a success. Other projects would not have been possible without similar efforts by Giovanni Galizia, Matthias Kohl, Elmar Lang and Frank Sengpiel.

It is hard to develop ideas and results without input from and discussions with other scientists, and even harder to form a book out of these results without the encouragement and the help of patient souls. I would like to thank Hauke and Ingo for reading and commenting on parts of the manuscript, and Ute Bauer for providing figures. Thank you also to Rita Almeida, Silvia Corchs, Gustavo Deco and Cristina Marchetti for giving me the feeling that this book might be worth existing and for their critical comments on editorial and philosophical aspects, including the cover design. Also, I wish to thank staff at Kluwer Academic Publishers for their help in the final stages of preparing this book. Finally, I thank my wife Beate for all her support and considerable patience in light of the many hours of extra work required to finish this book.

# Foreword

The complexity of the cerebral cortex demands appropriate tools for exploration of its function and its relationship to underlying anatomical wiring. This book presents a discussion of contributions made by computer science to both analysis of biological imaging data and construction of neural models of adult and developing visual cortex. The choice of visual cortex is a particularly fortunate one in that there is a wealth of anatomical, physiological and imaging data available concerning the primary visual cortex, area V1.

The author discusses how the tools of the computer scientist can be used to develop statistically sound methods of data analysis for optical imaging of the visual cortex. Since optical imaging is one of the most useful methods available for mapping the patterns of activity of very large ensembles of neurons, it is particularly important that the correct analytic tools be used. The modular organisation of different functions across the surface of visual cortex presents an ideal substrate for refining statistically appropriate approaches to image abstraction. Martin Stetter then goes on to discuss a variety of modelling approaches, based on feasible neural circuits. He describes possible neural substrates that might underlie the observed functional maps seen in area V1 and their larger scale implications. These models nicely illustrate the powerful contributions that computer science can make to understanding cortical function.

Martin Stetter has had considerable experience in the training of young computer scientists; his concern for accuracy and his interest in neurobiology has guided these students along valuable paths in applying computer technology to study of the cerebral cortex. The tools he discusses in this book are now essential elements in interdisciplinary cortical research as a whole, not just in the study of visual cortex, which he uses as a model system.

# Chapter 1

# INTRODUCTION

One of the central hypotheses in modern neuroscience is that all behavior is generated by the brain. It forms our perceptions from physical properties of our environment and controls our reactions and actions. It is the source of spontaneous thoughts and meanings, and it provides us with consciousness, imagination and emotions. Many generations of neuroscientists have devoted their careers to the goal of revealing the mechanisms, by which these complex aspects of behavior emerge from the anatomical and physiological properties of the brain. Although all these efforts have led to considerable progress in many disciplines of neuroscience, we are still far from even catching a glimpse on many of the principles of brain function.

One important reason for our obvious lack of success is the sheer complexity of our brain at many levels. Loosely speaking, the human brain forms an enormous network of one hundred billion nerve cells, which are densely interconnected by about $10^{15}$ synaptic connections. Each neuron and each synapse again represent complex dynamical systems with nonlinearities from the cellular down to the molecular level. In addition, the properties of the elements and the network are not fixed in time but can be changed by modulatory subsystems and synaptic plasticity. We are facing a giant nonlinear recurrent system, which progressively re-adjusts itself under the control of intrinsic states and the external environment. It seems fair to state that the human brain represents the most complex system in our known universe. How then can we address this apparently unsolvable problem of understanding the principles of brain function?

This problem can only be properly addressed by a multidisciplinary approach, by which two different research lines are tackled in parallel. Experimental disciplines including biology, medical sciences and psychology increase the pool of experimental data from various brain systems. This effort is complemented by techniques from mathematics, physics, engineering, computer science and related disciplines, which characterize relevant information in the data and extract basic principles and concrete mechanisms of brain operation. In addition, the common approach has to be interdisciplinary: It is not sufficient for the disciplines to work side by side in parallel. Instead different research lines must actively interact and guide each other. For example, the structures of theoretical models have to obey constraints which are dictated by experimental results, and conversely their testable predictions should guide the design of new experiments and should help formulating new and useful questions to ask.

The contribution of physics and computer science to this common undertaking is at least twofold. First of all, the discipline can yield techniques for processing experimental data in order to extract relevant features. Challenges for data analysis can come in many tasks including clustering, object segmentation, signal source separation, time series analysis and time series prediction, which can be treated using statistical data modeling techniques such as density estimation and function approximation. Secondly, these disciplines can provide computational models for the function and development of complex subsystems of the brain. One class of models treats the brain or some part of it as a nonlinear dynamical system, from which a simplified model is derived and theoretically analyzed. Usually, this type of models is based on the anatomical and physiological findings and aims at extracting operational principles of the considered system. These models form the category of bottom-up approaches. An alternative approach treats the brain itself as a signal processing and statistical inference machine. It can be asked which principle of operation or which design principle would enable the brain or a subsystem to optimally extract and represent relevant statistical structure in the environment or in a set of actions. Models of this class are top-down models.

We might ask, why it is beneficial for the scientific community to accept the challenge of revealing brain function. First of all, we could argue that it is most satisfying for a scientist to address one of the oldest questions of mankind: How can we do all these complex things? More importantly, however, for many classes of problems our brain is obviously by far the most powerful signal processing system we know. Gaining knowledge about its operational principles should provide us with a variety of fundamentally new and powerful signal processing algorithms, which can be formalized and transferred to technology and application.

Due to the complexity of the brain, we are forced to restrict ourselves to the exploration of adequate sub-systems. This approach follows a reductionist principle, which is not justified *per se*, because one of the most prominent features of the brain is its dense connectivity. Therefore, by modeling the operation of an isolated sub-system, we must at some point address the question of how its functional principles are altered by the connections it receives from and sends to other brain regions.

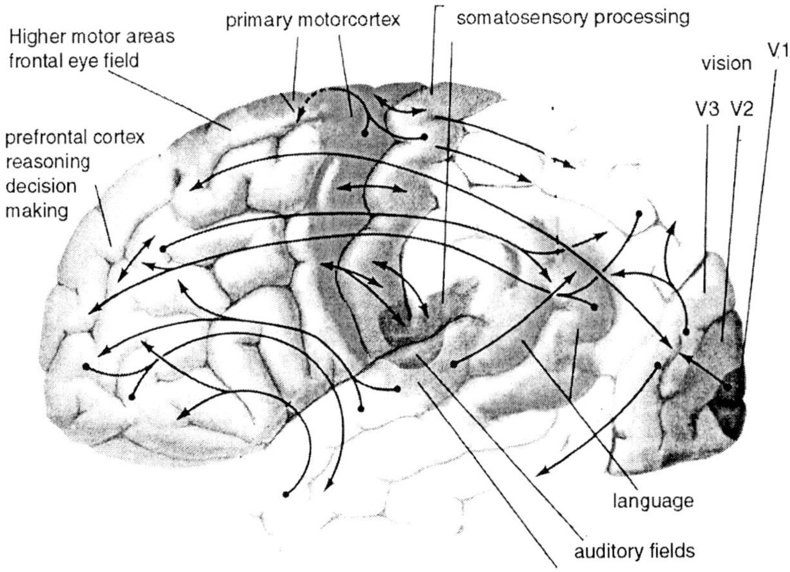


*Figure 1.1.* Illustration of a human brain, which is covered almost entirely by a laminar and strongly folded structure, the neocortex. Though being relatively uniform in structure, the neocortex contributes to a large variety of different signal processing tasks.

There are several reasons for considering the neocortex of higher mammals as an important subsystem of the brain. First, in most higher mammals the neocortex is a large and prominent structure, namely an about 2mm thick lamina of nerve cell bodies which covers large parts of the rest of the brain (figure 1.1). Its size already underlines its importance for higher brain function. Second, The neocortex can be functionally subdivided into a set of different areas (Van Essen et al., 1992), each of which, in the spirit of complexity reduction mentioned above, can be explored as a separate sub-system. Third, the anatomical structures of the cortex in different areas are very similar to each other (for an introduction see (Kandel et al., 1991)). This appears even more remarkable as we realize that the different cortical areas are devoted to completely different signal processing tasks including processing of visual, auditory and somatosensory signals, motor control, and even complex behavioral tasks such as reasoning, decision making and control of social integrity (figure 1.1; for an overview see (Damasio, 1996)). Loosely speaking, a cortical area appears as a universal signal processing module. Hence, by understanding the principles of operation for one area, we can hope to have gained a set of universally applicable signal processing strategies, which we can use either as a guide for understanding other cortical areas or for the design of a new class of innovative computer algorithms.

One cortical area, on which many laboratories concentrate their joint efforts is the primary visual cortex. The present book describes some contributions of modern computer science towards a better understanding of its general functional principles. However it cannot and does not even attempt to be a comprehensive on this issue. For example, we will not address the important fields of temporal coding in biological systems (Eckhorn et al., 1988; Gray et al., 1989) or in artificial neural networks (Maass and Bishop, 1998; Maass, 2000), but rather exemplify a selection of recent approaches towards the an understanding of spatial visual representation.

The book can be logically divided in three parts. The first part comprises chapters 2 and 3, in which I will summarize some general and important aspects of how biological neurons process information, and how they are interconnected to form the early visual pathway in macaque monkeys. We will find that one important feature of cortical signal processing is encoding of information by the activity of large neuron populations.

The second part is devoted to methods for measuring the activity patterns in large populations of cortical neurons, which are needed as experimental inputs for computational models. In chapter 4 we will learn about the principles of optical imaging of brain activity. We shall consider some of its limitations and formulate possibilities for improvements. Chapters 5 to 8 address the question of how we can reliably infer neuronal activity patterns from measured imaging signals. This task will first be formulated as a blind source separation problem, for which we then summarize a set of algorithms together with some applications to functional imaging data.

The third part will introduce computational approaches towards a quantitative understanding of cortical functional principles. Chapter 9 will focus on top-down approaches, which try to understand aspects of cortical function as the implementation of a design principle or optimality criterion of signal processing. In chapter 10, these models will be contrasted by mean-field models of cortical function as bottom-up approaches, which are based on the description of neuronal population activities and anatomical knowledge about neuronal wiring patterns. These models will address the question how cortical function evolves as a consequence of its anatomical construction principles, and which of these construction principles are crucial for the generation of its functional principles. Examples will include representation of orientation and contrast, but also the processing of more complex stimuli and the use of visual context for image segmentation.

# Chapter 2

# NEURONS AND NEURONAL SIGNAL PROPAGATION

The human brain contains approximately $10^{12}$ to $10^{13}$ cells. About $10^{11}$ cells out of this pool can be classified as nerve cells or neurons. Although representing only 2–10 % of all brain cells, it is widely accepted that the neurons and their mutual synaptic connections form the anatomical substrate for the powerful computational abilities of the brain. The remaining cells, many of which are called glial cells, are devoted to important auxiliary tasks: They ensure optimal conditions for neuronal signal transmission by preserving important ionic concentration levels and insulating neurons from each other, they accelerate neural signals by myelinization, they shield the brain against toxic substances in the blood by forming the *brain-blood barrier* and they contribute to the nutrition of neurons. In chapter 4, we will discuss a potential key role of glial cells for the optical measurement of nerve cell activity. Nevertheless, the activities of glial cells are thought to be only of secondary importance for understanding the principles of neuronal signal processing.

In the this introductory chapter, we will summarize some important aspects of the morphology of nerve cells and of the transmission of neuronal signals. After a brief overview in section 1, sections 2 – 5 will focus on different aspects of signal propagation in neurons. Further general reading for neurobiological issues can be found in (Shepherd, 1988; Kandel et al., 1991; Dowling, 1992), and for models of neuronal signaling in (Jack et al., 1975; Bower and Beeman, 1994; Koch and Segev, 1998).

## 1. Overview

Neurons are cells, which are specialized to serve their task of signal transmission and signal processing. Figure 2.1 summarizes some reconstructions of different cell types. These neurons belong to the class of multipolar nerve cells, on which we will concentrate our further description. Though there is a

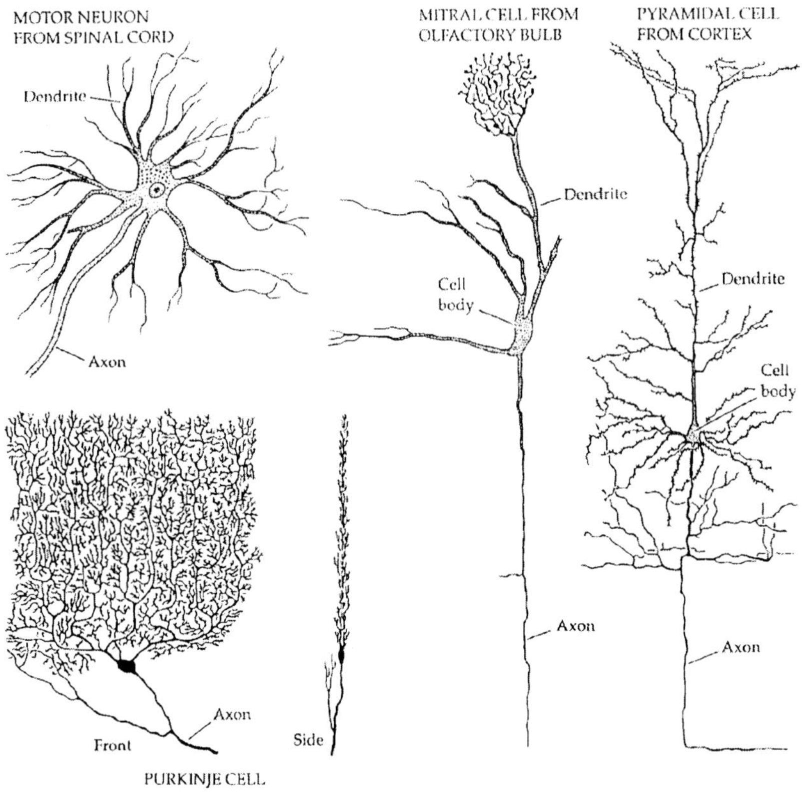


*Figure 2.1.* Illustrations of different neuronal morphologies.

variety of different morphological shapes, all multipolar neurons have a basic anatomical structure in common, which is schematically illustrated in figure 2.2. The central part of a neuron is its cell body or soma, which contains the nucleus and other important cell compartments. Attached to the soma we find many tree-like and strongly branching structures, the dendrites. In addition, each neuron has a single further process. It is called the axon and its connection with the cell body is referred to as the axon hillock. Axons can become very long, they can range from fractions of a millimeter up to several meters for motoneurons of the spinal chord. Therefore, many axons in the central nervous system are myelinized by glial cells which increases the conduction velocity. Each axon branches into many axonal fibers, on which we can find synaptic terminals. These terminals form a synaptic connection with the dendrite, the soma or the axon hillock of other neurons, which are called "postsynaptic" neurons. On average, every axon forms about 10 000 synapses with postsynaptic

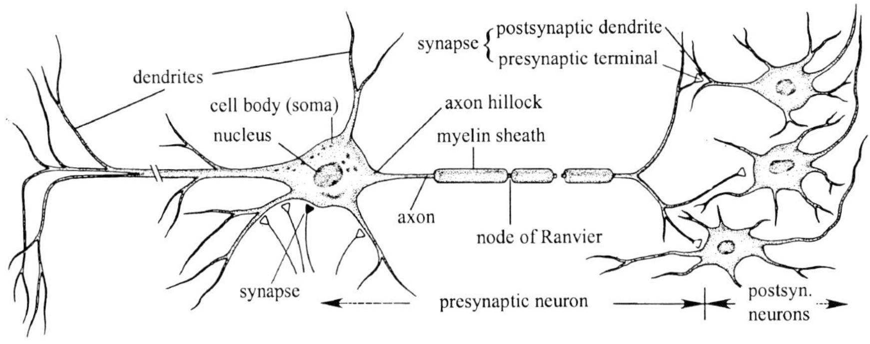


*Figure 2.2.* Schematic illustration of a neuron. A neuron consists of dendrites, the cell body and a single axon fiber which forms synaptic connections with other, "postsynaptic" neurons (adapted from (Kandel et al., 1991)).

neurons, and each cell receives roughly the same number of connections from other neurons. The complex shape of the neuron is defined by the cell membrane, which separates its interior from the external milieu. The membrane is kept in place by the molecular cytoskeleton and by adhesion to neighboring cells.

Neuronal signals are changes of the resting electrical potential across the cell membrane, which at rest is maintained at about $V_R = -70$ mV [1]. By their function for the transmission and transformation of these changes, the morphological parts of a neuron can be grouped into four major functional components: an input component, an integration component, a conduction component and an output component (figure 2.3).

The dendrites and the cell body are densely covered with synaptic terminals from many other neurons, which evoke changes in the membrane potential. These changes are called postsynaptic potentials (PSPs). Dendrites are the input component of the neuron (figure 2.3a).

At the axon hillock, many PSPs are integrated to form a total synaptic potential, and are transformed to a sequence of sharp pulse-like potential changes, the action potentials or spikes (figure 2.3b). In a simplified view this transformation can be seen as a threshold operation: Each time the total membrane potential exceeds a threshold voltage $V_{\text{th}}$, a spike is generated and the membrane potential is reset to its resting value. This operation can be viewed as a summation with a subsequent analog-to-digital transform. The axon hillock represents the integration component. The resulting sequence of spikes is conducted through

[1]The exact value depends on the species, temperature and other factors

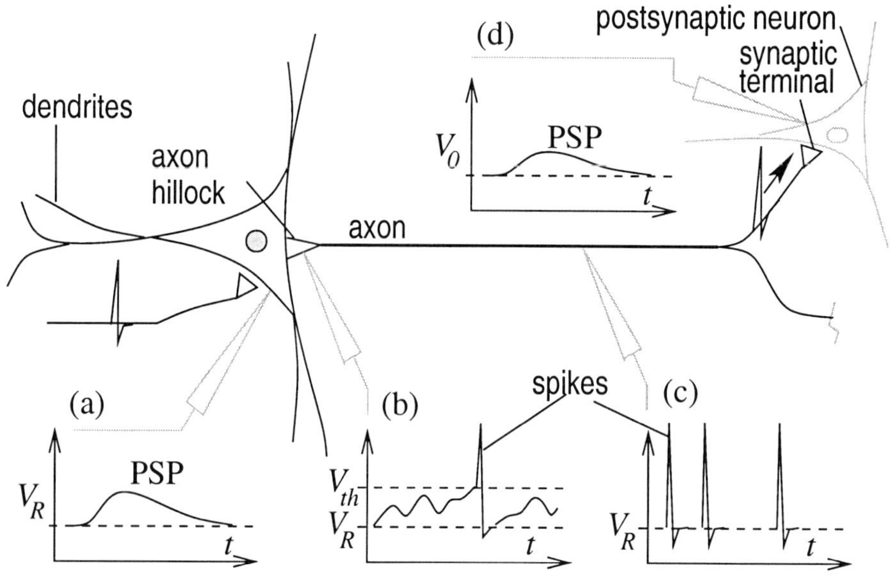


*Figure 2.3.* Simplified overview over neuronal signal flow. **(a)** The dendrites collect postsynaptic potentials (PSPs). **(b)** At the axon hillock, PSPs are summed and spikes are generated by a thresholding operation. **(c)** The spikes are conducted along the axon to the synaptic terminals, where they evoke PSPs in postsynaptic neurons **(d)**.

the axon (figure 2.3c), which the conduction component, and is distributed to all synaptic terminals, which form the output component. Each time a spike arrives through the axon, the synapse causes a smooth postsynaptic potential in the subsequent neuron (figure 2.3d). This operation can be understood as a digital-to-analog transform. Chemical synapses transmit information only along one direction. The neuron on the axon side, which emits the signal, is called the presynaptic neuron, the corresponding synaptic compartment is the presynaptic terminal. Likewise, the recipient neuron is referred to as postsynaptic neuron.

Based on this overview, the following paragraphs will provide some more details about the different stages of the signal processing in neurons.

## 2. The Resting Potential

The main constituent of the cell membrane is a phospholipide bilayer, which is impermeable for water and ions. It separates the extracellular from the intracellular fluid, both of which consist of water and various types of ions and organic molecules. Embedded into the cell membrane we find specialized proteins, many of which act as specific channels for one or a few types of ions.

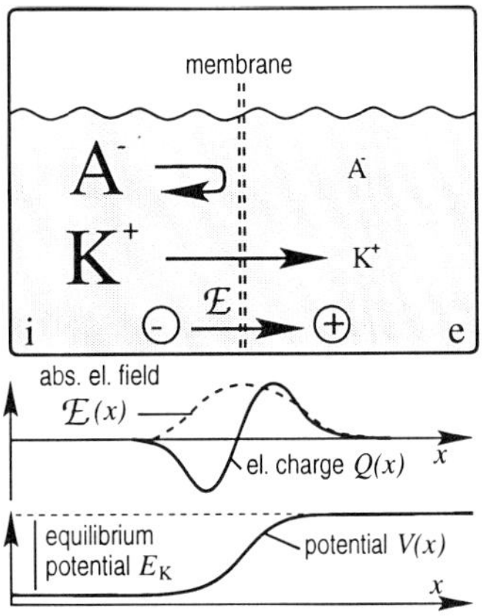


(a)

| Ion | Internal (mM) | External (mM) | Equilib.-Pot. (mV) |
|---|---|---|---|
| $K^+$ | 400 | 20 | $E_{\mathrm{K}}$: -75 |
| $Na^+$ | 50 | 440 | $E_{\mathrm{Na}}$: +55 |
| $Ca^{2+}$ | 0.4 | 10 | $E_{\mathrm{Ca}}$: +41 |
| $Cl^-$ | 52 | 560 | $E_{\mathrm{Cl}}$: -60 |

(b)

*Figure 2.4.* **(a)** Generation of the equilibrium potential for potassium by a concentration gradient and a selectively permeable cell membrane (for explanation see text). **(b)** Concentrations of important ions in the intracellular and the extracellular space and their resulting equilibrium potentials (after (Kandel et al., 1991)).

The resting potential of a neuron is generated by two effects: $(i)$ Several ion types show a concentration gradient along the membrane, which is actively maintained by the cells under energy consumption. $(ii)$ The membrane contains passive ion channels, which make it partially and selectively permeable for some ions.

The most important ion types for the resting potential are potassium ions $K^+$ and large organic anions (denoted by $A^-$). Figure 2.4a shows a simplified scenario, in which a piece of membrane (vertical dashed line) separates the cytosol (left) from the extracellular space (right). The concentration of potassium ions within the cell is approximately 20 fold higher than its concentration outside the cell: $[K^+]_i \approx 20 \times [K^+]_e$[2]. The fluid is electrically neutralized by organic anions $A^-$.

Because the membrane is selectively permeable for potassium, these ions diffuse outside the cell along their own concentration gradient, and leave the non-permeating anions behind (figure 2.4a, top). This process causes a surplus of positive charges outside and a surplus of negative charges inside the cell, which attract each other and concentrate at the membrane. The separated charges evoke an electrical field and a corresponding electrical potential, which increasingly prevent further $K^+$ ions from leaving the cell. When the chemical and electrical forces cancel, the system has reached its equilibrium. Figure 2.4a bottom schematically illustrates, how the electrical charge $Q$, the strength of the

[2]The square brackets denote the concentration of an ion type

electrical field E and the electrical potential $V$ are distributed across the membrane. The difference $E_{\mathrm{K}} = V_i - V_e$ between the potential inside and outside is the equilibrium potential of potassium, which is negative. In summary, the membrane acts as the dielectricum of a capacitor which separates the charges inside and outside the cell from each other.

The equilibrium potential for potassium can be quantified by equating the electrical force and the chemical force acing on the ions. This yields the Nernst-equation

$$E_{\mathrm{K}} = \frac{RT}{ZF} \ln \frac{[\mathrm{K}^+]_e}{[\mathrm{K}^+]_i} \approx 60 \log_{10} \frac{[\mathrm{K}^+]_e}{[\mathrm{K}^+]_i} \ \mathrm{mV}, \tag{2.1}$$

in which $R$ is the gas constant, $T$ the absolute temperature, $Z = +1$ the valence of potassium and $F$ the Faraday constant. In the second equation, the prior factor of the logarithm has been calculated for room temperature (25 deg Celsius) and the logarithm has been changed to base 10.

However, potassium is not the only ion type which is present in living tissue. The table in figure 2.4b provides the intracellular and extracellular concentrations of further important ion types for the giant squid axon. Based on these concentration conditions, each ion has its own equilibrium potential. It can be obtained by insertion of the respective concentration values into the Nernst equation eq. (2.1). The equilibrium potentials for some important ions are listed in the right column of figure 2.4b. Notice that sodium and calcium have a positive equilibrium potential, their concentration gradient drives them into the cell.

The equilibrium potential of an ion represents the membrane voltage we would measure if the membrane were permeable only for that ion. Actually, however, the membrane is to some degree permeable for many ions (figure 2.4b), and all of them together determine the membrane potential $V_m$ of the neuron. If the permeability for a given ion type is very low, it cannot contribute much to a shift of charges across the membrane and thus does not strongly affect the membrane potential. Conversely, ions with a high permeability dominate the value of the membrane potential. This relationship is reflected by the Goldman equation,

$$V_m = \frac{RT}{F} \ln \frac{P_{\mathrm{K}}[\mathrm{K}^+]_e + P_{\mathrm{Na}}[\mathrm{Na}^+]_e + P_{\mathrm{Cl}}[\mathrm{Cl}^-]_i}{P_{\mathrm{K}}[\mathrm{K}^+]_i + P_{\mathrm{Na}}[\mathrm{Na}^+]_i + P_{\mathrm{Cl}}[\mathrm{Cl}^-]_e} \tag{2.2}$$

in which $P_{\mathrm{K}}$, $P_{\mathrm{Na}}$ and $P_{\mathrm{Cl}}$ denote the permeabilities of the cell membrane for potassium, sodium and chloride, respectively. Other permeabilities have been neglected. If any of the permeabilities is much higher than the others, the Goldman equation will approximately reduce to the Nernst equation for the corresponding ion. The value of the membrane potential will be dictated by this ion and will come close to the respective equilibrium potential.

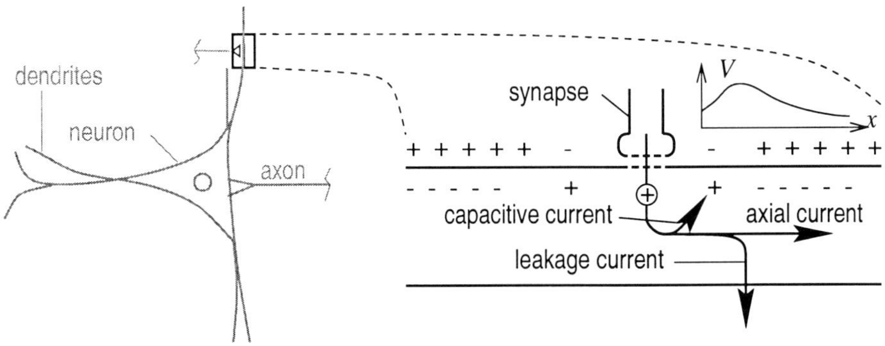


*Figure 2.5.* Major currents that contribute to the passive signal propagation in dendrites. An injected synaptic current splits into a capacitive current, an axial current and a leakage current. The example shows an excitatory synapse, in which the capacitive current causes a depolarization of the membrane.

At rest, the membrane permeability is strongest for potassium, $P_K \gg P_{Na}$, $P_{Cl}$, and consequently the membrane potential amounts to a value only sightly above its equilibrium potential: $V_m \approx -70$ mV. Under these conditions, the membrane potential is called the *resting potential* $V_R$ of the cell. In section 4 we will see that during an action potential, the membrane permeability is highest for $Na^+$ and the membrane potential is dominated by sodium.

## 3. Passive Signal Propagation in Dendrites

The signals that are processed by neurons are deviations of the membrane potential from its resting value. As mentioned previously, these perturbations are usually caused by presynaptic terminals, which evoke a current injection into (or out of) the postsynaptic neuron when they are activated (cf section 5). The evoked perturbations or postsynaptic potentials then propagate through the dendrites towards the soma, where they are integrated with other PSPs to form the total synaptic potential. Dendrites are complex structures and signal propagation may be accompanied by active calcium-dependent processes (Mel, 1993; Johnston et al., 1996; Destexhe et al., 1996; Schutter, 1999), which renders a comprehensive treatment of this issue beyond the scope of this introduction. Dendritic integration of signals has been addressed experimentally, (Segev and Parnas, 1983; Fromherz and Vetter, 1992; Fromherz and Müller, 1994; Klingler and Fromherz, 1995) by analytical model treatment (Rall, 1964; Rall, 1967) and by compartmental modeling (Bressloff and Taylor, 1993a; Bressloff and Taylor, 1993b; Bressloff, 1994; Segev, 1995; Schutter, 1999). At this point, we restrict ourselves to a a brief consideration of what are the principles of passive signal propagation in dendrites. Let us consider a small piece of a dendrite,

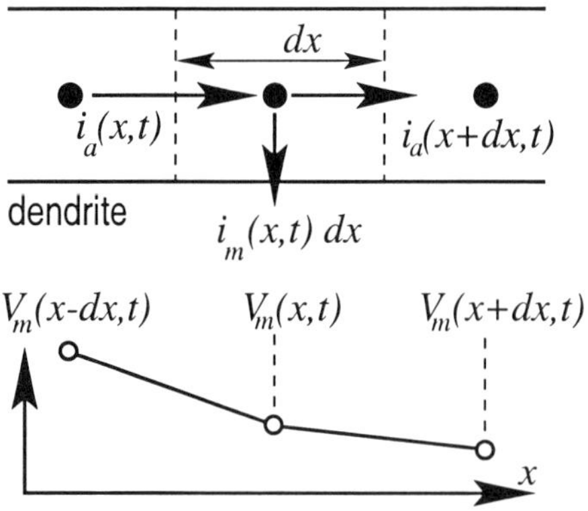


*Figure 2.6* Currents and voltages during signal propagation along an idealized cylindrical dendrite. In each segment the currents must cancel. Hence the membrane current is provided by changes in the axial currents, which in turn are driven by differences in membrane potential.

which we assume to be very thin and very long, and which is contacted by a presynaptic terminal (figure 2.5). When the synapse is driven by a presynaptic spike, it causes the injection of a small current pulse into the dendrite. This current splits into three basic components: $(a)$ The injected ions change the charge distribution across the membrane near the synapse and thereby locally change the membrane potential. The corresponding current is the capacitive current $i_c$. $(b)$ Ions diffuse along the dendrite and form the axial current $i_a$. The axial current is responsible for the propagation of the depolarizing perturbation. $(c)$ Ions leak through passive channels in the membrane and form the leakage current, $i_l$, which drives the membrane potential back to its resting level. The leakage current is thought to be mediated by chloride ions.

Now we want to derive an expression, which specifies how a deviation of the membrane voltage from its resting level propagates through this idealized dendrite. For this it is convenient to think of the dendrite as a sequence of short segments, each with infinitesimal length $dx$ (figure 2.6). Within each dendrite segment $x$ and at each instant of time $t$, the total amount of charge is preserved: The sum of incoming currents must be equal to the sum of outward currents. Equivalently, the axial current $i_a(x,t)$ which enters the segment, but does not leave it again axially as $i_a(x+dx,t)$, has to leave it as resistive or capacitive membrane current $i_m(x,t)dx$:

$$i_m(x,t)dx + i_a(x+dx,t) = i_a(x,t); \qquad -\frac{di_a(x,t)}{dx} = i_m(x,t) \tag{2.3}$$

Each of the axial currents is driven by the voltage between itself and its neighboring segment. This yields

$$i_a(x,t) = \frac{V_m(x-dx,t) - V_m(x,t)}{r_a dx} = -g_a \frac{dV_m(x,t)}{dx}, \tag{2.4}$$

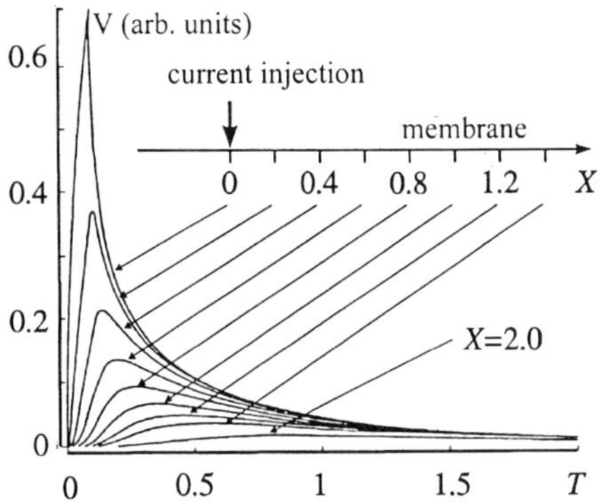


*Figure 2.7* Solution of the cable equation for an infinite cylindrical dendrite in response to a brief current injection at $T = X = 0$. $X = x/\lambda$ and $T = t/\tau_m$ are reduced length and time variables. With increasing distance from the synapse, the PSP becomes delayed, weaker and broader ((Stetter, 1994), adapted from (Jack et al., 1975)).

where $g_a$ (mS cm) denotes the conductance per unit length along the axial direction of the dendrite. Finally, the membrane current can consist of many components, but at least includes the capacitive current and the leakage current, which may be carried by several types of ions. We obtain

$$i_m(x,t) = c_m \frac{dV_m(x,t)}{dt} + g_m(V_m(x,t) - V_R), \qquad (2.5)$$

where $g_m$ (mS/cm) and $c_m$ ($\mu$F/cm) represent the membrane conductance and the membrane capacitance, respectively, per unit length of the dendrite. By deriving eq. (2.4) with respect to $x$, inserting eqs. (2.4) and (2.5) into (2.3) and dividing by $g_m$ we arrive at the cable equation

$$\lambda^2 \frac{d^2V_m(x,t)}{dx^2} - \tau_m \frac{dV_m(x,t)}{dt}. - (V_m(x,t) - V_R) = 0 \qquad (2.6)$$

The cable equation describes, how the membrane voltage of a long cylindrical dendrite behaves in space and time, when the membrane voltage is perturbed from its resting level. The first term in (2.6) is a diffusion term which specifies how the ions propagate along the dendrite. It corresponds to the axial current in figure 2.5. The second term describes how the membrane is recharged and can be traced back to the capacitive current in figure 2.5. The third term, finally, is the leakage term which drives the membrane potential back to zero. The quantitative properties of the dendrites are specified by the length constant, $\lambda = (g_a/g_m)^{1/2}$ and the time constant $\tau_m = c_m/g_m$. The length constant specifies how far a local perturbation can propagate along the dendrite and ranges around $0.3 - 1$ mm. The time constant determines how long a perturbation of the membrane potential is typically preserved. It assumes values around $5 - 100$ ms.

The cable equation represents a second order partial differential equation, for which a general solution even in the idealized case of a infinite cylindrical dendrite is difficult to find. However, the propagation of a transient voltage change in response to a short localized current pulse can be well analyzed by this equation. Figure 2.7 shows, how the voltage over time behaves at different locations at the dendrite . The variables in capital letters denote effective time and space, $X = x/\lambda$ and $T = t/\tau_m$. Along their way towards the soma, the postsynaptic potential decreases in amplitude and becomes increasingly broader. Hence, with increasing distance from the soma, the influence of a synaptic current on the somatic potential as well as its temporal localization decreases.

We conclude that signal propagation in passive dendrites can be understood as a linear process, namely as a diffusion-based spread of charge in a linear system. Because of these two properties, this mode of dendritic signaling is called passive propagation. Passively propagated PSPs decay exponentially and broaden along their way through the dendrite. At the soma, they show a variety of strengths and degrees of temporal localization. The total potential at the soma or the axon hillock results from a complex spatial and temporal summation of usually tens to hundreds of individual postsynaptic potentials.

## 4. Active Propagation of Spikes

The characteristics of signal propagation at the axon differs dramatically from that in dendrites. Starting from the axon hillock and the axon initial segment, each neuron fires a sequence of short and sharp pulse-like changes in the membrane potential, the action potentials or spikes. Their temporal pattern depends in a complex way on the summed PSP of the dendrite. The spike sequence or spike train encodes and processes the summed dendritic potential. Further reading on experimental and theoretical approaches to excitable membranes includes (Fitzhugh, 1960; Fitzhugh, 1961; Jack et al., 1975; Abbott and Kepler, 1990; Ermentrout, 1996; Ermentrout and Kopell, 1998).

The thick solid line in figure 2.8a illustrates, how the membrane voltage changes in time during an action potential. As soon as the membrane voltage passes a threshold value, it increases very fast up to a value of about $+50$ mV and afterwards decreases even below the resting potential, which it reaches again after a few milliseconds. In mammals, the whole spike takes roughly one millisecond. Immediately after the spike it is harder than usual for the neuron to fire a new spike. This period is called the refractory period of the neuron.

At its peak value, the membrane voltage is close to the equilibrium potential of $Na^+$. This indicates, that during that period the membrane permeability for sodium is much higher than the others, $P_{Na} \gg P_K, P_{Cl}, \ldots$ and dominates the behavior of the membrane potential. Hence, the activation of additional sodium channels is likely to contribute to the initiation of a spike. After the spike, the membrane potential is even closer than usual to the equilibrium potential

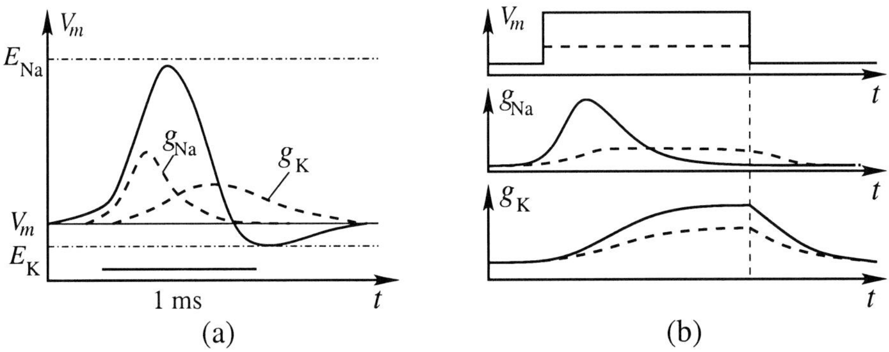


*Figure 2.8.* **(a)** Shape of an action potential (solid line) and changes of the sodium and potassium conductances (dashed lines) during an action potential. **(b)** Voltage-dependence of $g_{Na}$ and $g_K$. Above a minimum membrane depolarization (threshold voltage, top), voltage-gated sodium channels open fast and inactivate automatically (center). Potassium channels activate slower and do not inactivate (bottom).

of $K^+$. Consequently, the activation of additional potassium channels should participate in the repolarization of the membrane.

The mechanism of spike generation has been revealed for the squid giant axon by (Hodgkin, 1948) and (Hodgkin and Huxley, 1952). By an electronic feedback circuit, which is known as *voltage clamp*, it became possible to keep the membrane voltage fixed at arbitrary values and simultaneously measure the membrane conductances for either potassium or sodium ions (at each experiment, one of the ion types was replaced by a substitute ion). They found that in contrast to the dendritic membrane, the axonal membrane is characterized by ionic conductances which depend both on time and on the current membrane voltage. They are related to voltage-gated ion channels in the membrane. Figure 2.8b summarizes the nonlinear behavior of the sodium and potassium conductances. If the membrane voltage increases only slightly above the resting potential, both conductances do not change very much. However, as soon as the change in membrane voltage exceeds a certain threshold, the sodium conductance $g_{Na}$ shoots up very fast to a high value and afterwards drops back to a low level (inactivation), even if the membrane is still depolarized. In contrast, the potassium conductance $g_K$ increases slower in response to depolarization and does not inactivate. Finally, a small, voltage independent leakage current was found, which is mediated by chloride and other ions (not shown).

Using this knowledge about the voltage-dependent conductances, one can understand a simplified version of spike formation: As soon as the membrane voltage at the axon hillock increases above a threshold value, voltage-gated $Na^+$ channels activate, cause an influx of $Na^+$ ions which further depolarize the membrane and cause even more $Na^+$ channels to open. A positive feedback loop is initiated, after which the membrane is highly permeable for $Na^+$. Shortly after this moment, the sodium channels inactivate automatically and stop the sodium flux, while at the same time the slower potassium channels open. Potassium leaves the cell and drives the membrane potential towards its equilibrium potential, which lies even below the resting potential. This means that the effect of an earlier membrane depolarization before the spike is reset. Finally, a slower process brings both the membrane potential and the potassium channels back to their resting state. The dashed lines in figure 2.8a illustrate the changes of the conductances during a spike. During a few milliseconds after spike initiation, the $Na^+$ channels are still closed due to their own inactivation process and cause the refractory period of the neuron.

Once a spike is initiated, it propagates along the axon towards the axon terminals. The propagation results from the fact that the spike represents a zone of strong membrane depolarization. This zone depolarizes the adjacent axonal segment, which in turn starts to initiate a spike using its own voltage-gated ion channels. This effect happens again and again and causes a continuous propagation of the action potential along the axon. The spike can only propagate away from the axon hillock, because the membrane over which it has already swept is refractory and cannot be excited again for a while.

A few points are worth noticing. Action potentials are generated by a nonlinear process, they preserve their shape during propagation along the axon. This has two consequences: First, the detailed shape of the action potential probably does not carry much information. Instead, the timing of a spike relative to preceding and successive spikes on the same axon or relative to spikes of other neurons seem to be important variables for encoding of information. This hypothesis gave rise to integrate-and-fire neuron models. Second, the various kinds of ion movements across the membrane evoke changes in cellular volumes and shapes. The latter two mechanisms, activity-dependent metabolism and cell swelling, represent important bases for the optical detection of neuronal activity, which is treated in chapter 4.

If we inject a stationary depolarizing current into the cell, its membrane voltage increases, passes the threshold voltage and causes a spike. The more current we inject, the faster the threshold is reached, and the shorter the interspike interval gets. The relationship between the injected stationary current and the resulting frequency of spikes is called the current-frequency relationship of a cell. Figure 2.9a shows a current-frequency relationships for two types of neurons, pyramidal cells and stellate cells, in the primary visual cortex of cats. The dashed lines are linear fits to the nonzero parts of the two relationships, which describe the data very well. Hence, the integrative part of the neuron (the translation from a stationary current to spike frequency at the axon hillock) can be described by a simple rectifying and piecewise linear function, at least

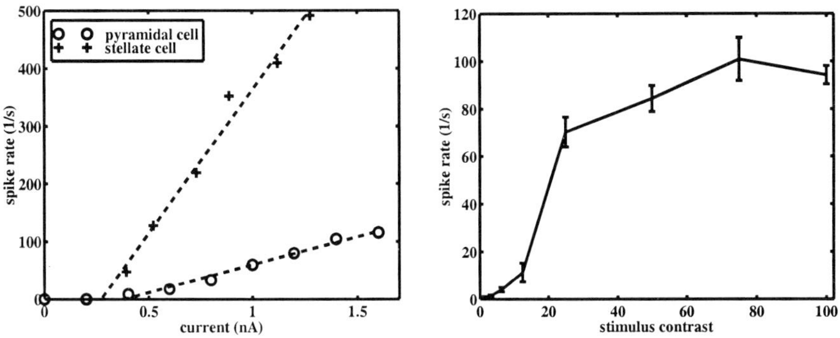


(a) (b)

*Figure 2.9.* **(a)** Current-frequency relationships for a pyramidal neuron (circles) and a stellate cell (crosses). The dashed lines are linear fits to the nonzero parts of the data (from (Azouz et al., 1997) and (Ahmed et al., 1998)). **(b)** Response of a pyramidal neuron in macaque visual cortex to the contrast of visual stimulation (Levitt, unpublished result). The cell shows strong contrast saturation.

up to a maximum frequency which is determined by the refractory period of the neuron. However, this does not imply that signal processing in networks of such neurons behaves linearly. Figure 2.9b shows, how the response of an orientation-selective neuron in the primary visual cortex of a macaque varies with the contrast of the stimulus. The neuron shows a strongly nonlinear response to the stimulus, including a saturation for high contrast levels. Neurons in cat visual cortex behave similarly (Sclar and Freeman, 1982). This nonlinear behavior is probably caused by non-stationarity of realistic synaptic inputs to nerve cells and by the fact that each neuron acts as part of a large neuronal network, which processes the visual stimulus. We will treat computational models for signal processing by neural networks in chapters 9 and 10.

## 5. Synaptic Transmission and Plasticity

Signals between neurons are exchanged via synapses. Two major classes of synapses can be found: Electrical synapses or gap junctions are fast electrical connections between cells, through which currents can be exchanged bidirectionally. In contrast, chemical synapses establish unidirectional connections, which use chemical transmission of signals. Chemical synapses are the most frequent and a very important type of synapses in the central nervous system. In the following, we will concentrate on a short review of synaptic morphology, function and plasticity of chemical synapses and will drop the expression 'chemical'.

Figure 2.10a sketches the asymmetric setup of a synapse. It consists of a small presynaptic terminal with a diameter of about 0.5-1 $\mu$m (top part) and a postsynaptic compartment (lower part). In the presynaptic part we find a num-

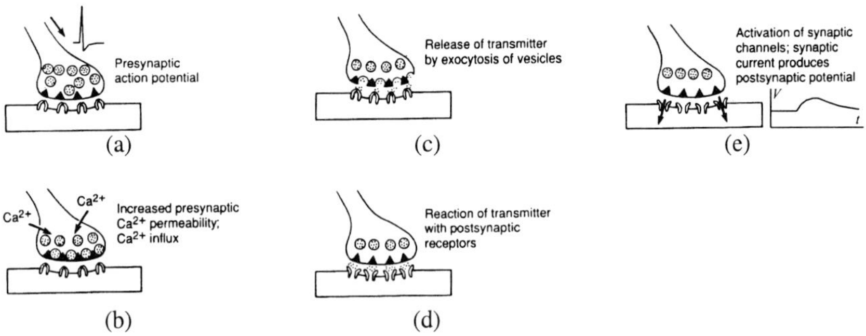


*Figure 2.10.* Principle of chemical synaptic transmission: A presynaptic action potential causes the release of neurotransmitter, which diffuses through the synaptic cleft, activates ion-channels in the postsynaptic membrane which enable an ion flux across the membrane (adapted from (Kandel et al., 1991)).

ber of small membrane vesicles, which contain neurotransmitter molecules and a set of release sites (black triangles) for these vesicles. Also, voltage-gated calcium channels are embedded in the presynaptic membrane. The postsynaptic membrane contains specialized ligand-gated ion channels, which have a specific receptor for the neurotransmitter molecules and are opened by a chemical binding to such a molecule. The procedure of synaptic transmission is summarized in figure 2.10. When a spike travels along an axon and arrives at the presynaptic terminal (figure 2.10a), it strongly depolarizes the presynaptic membrane. This activates the $Ca^{2+}$ channels, calcium flows along its gradient into the cell (figure 2.10b) and causes neurotransmitter vesicles to fuse with the cell membrane (figure 2.10c). The neurotransmitter molecules diffuse through the synaptic cleft, which is approximately 10 nm wide, and shortly after their release bind to the receptors of the postsynaptic ion channels (figure 2.10d). The channels are activated, the postsynaptic membrane permeability for certain ions increases, and the resulting ion flux changes the membrane potential (figure 2.10e) – a postsynaptic potential has been generated. Due to this cascade of steps, the onset of the PSP has a short delay with respect to the presynaptic spike. Also, it is less sharply peaked and in the mammalian brain extends to about 5-100 ms over time.

We have seen that current injection into the postsynaptic membrane is achieved by a change in electrical conductance of the postsynaptic membrane. Which change in the membrane voltage is evoked by a given synapse, depends on the combination of ion types for which it becomes permeable. If a synapse were only permeable for a given ion type, it would drive the membrane potential towards the equilibrium potential for that ion (cf. section 2), because the high permeability for that ion during synaptic transmission would let it dominate the right hand side of the Goldman equation (2.2). Several types of synapses,

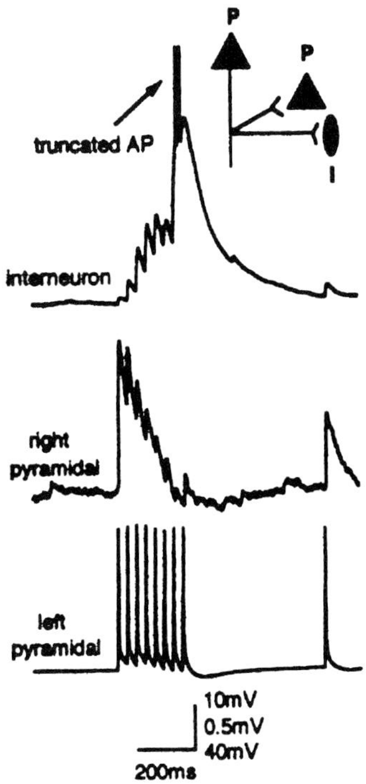


*Figure 2.11* Short term synaptic plasticity of excitatory synapses: The presynaptic spike train (bottom) drives a fast depressing synapse (middle) and a fast facilitating synapse (top) on the axon of the same presynaptic neuron (from (Markram et al., 1998)).

however, become permeable for a combination of ions. In this case, the synapse has its own equilibrium potential, which is called reversal potential.

A lacannot transfer the file rge fraction of synapses in the brain use glutamate as neurotransmitter. The ion-channels of glutamatergic synapses become permeable for both $Na^+$ and $K^+$ and have a reversal potential of about 0 mV. If such a synapse is activated, it depolarizes the membrane [3]. Synapses with this property are called excitatory synapses, because they drive the membrane voltage towards the firing threshold for a spike. They cause excitatory postsynaptic potentials (EPSPs) with a positive amplitude between $0.2 - 2$ mV at the soma. Glutamatergic ion channels can be subdivided into two classes: AMPA-receptors and NMDA-receptors. AMPA-receptors evoke strong and fast ($\approx$ 5 ms) EPSPs, whereas the contribution of NMDA-receptors is usually much weaker and slower ($\approx$ 100 ms). However, NMDA-receptors are both transmitter and voltage gated – they need a pairing of presynaptic transmitter release and a postsynaptic depolarization for activation. Also, they are permeable for $Ca^{2+}$, which is thought to be involved in the initiation of synaptic plasticity.

[3] unless the membrane voltage were already positive, which is usually not the case outside action potentials

A second important type of synapses use gamma-amino-butyric-acid (GABA) as neurotransmitter and their corresponding ion-channels are selective for $Cl^-$. Because in mammals the equilibrium potential of chloride lies below the resting potential of neurons, active GABAergic synapses hyperpolarize the membrane. Synapses with this property are called inhibitory synapses, because they drive the membrane voltage away from the firing threshold for a spike. The resulting inhibitory postsynaptic potentials (IPSPs) are much smaller in amplitude than the EPSPs, because of the proximity of the membrane potential and the reversal potential of the synapses. Nevertheless, inhibitory synapses can be very efficient even without causing a PSP by themselves: By increasing the membrane conductance, they establish a short-circuit through the membrane which can remove already existing EPSPs very effectively. This mechanism of inhibition is called shunting inhibition.

In response to a presynaptic spike, a synapse causes a PSP with a given amplitude. This amplitude represents one simple possibility to define the efficiency of transmission for that synapse, which is referred to as its synaptic efficacy or synaptic strength. The synaptic efficacy is in general different for each synapse. Theoretical studies on artificial neural networks (see e.g. (Koch and Segev, 1998) for an overview) suggest that the set of synaptic strengths in a network of neurons determines its mode of operation. However, the strength of a synapse is not fixed over time but can vary in an activity-dependent way over various time scales. This class of phenomena is called synaptic plasticity. Because this implies, that the mode of operation of a biological network can change in an activity-dependent manner, synaptic plasticity is believed to underly both fast adaptation processes (Markram and Tsodyks, 1996) and slower learning and memory mechanisms (Bliss and Collingridge, 1993).

Figure 2.11 demonstrates different types of short term synaptic plasticity, all of which act at a time scale from 0.2 s up to a few seconds. The measurements shows simultaneous recordings from three neurons in a slice of rat somatosensory cortex, one pyramidal neuron and two different postsynaptic neurons, to which it makes a synaptic contact. The lower trace shows action potentials that are generated by electrical stimulation in the presynaptic pyramidal cell, whereas the middle and top traces display the summed PSPs in the two postsynaptic neurons. Each little peak corresponds to one PSP which arose from a single spike. In the middle trace, PSP amplitudes strongly decrease over about 200 ms, they show fast synaptic depression. which partly recovers after approximately 0.5 s. The top trace, in contrast, shows fast synaptic facilitation with similar temporal characteristics.

Synapses can also undergo activity-dependent changes in strength which are persistent over long periods of time. If in response to a single presynaptic burst of activity the PSP evoked by a synapse becomes and remains stronger for at least some hours, the effect is called Long-Term Potentiation (LTP). Similarly, if activity diminishes the PSP for a long time, Long-Term Depression (LTD) has occurred (Linden and Connor, 1995). In many cases, LTD and LTP have been shown to require a pairing of presynaptic spiking activity and postsynaptic depolarization, which is in turn a sign of postsynaptic activity (for a review see (Brown et al., 1990)). In this case, synapses change their strengths based on the coincidence of pre- and postsynaptic activity and thus can detect and amplify correlations in firing. This class of synaptic plasticity is called Hebbian plasticity (Hebb, 1949). Synapses with NMDA receptors are discussed as a molecular substrate for Hebbian synaptic plasticity (Daw et al., 1993; Schutter and Bower, 1993), because these receptors cause an influx of calcium selectively in the case of paired pre- and postsynaptic activities.

Recent studies have examined different situations, under which LTP and LTD are observed. If synaptic plasticity is evoked by strong bursts of spikes, LTP has been observed, if the postsynaptic cell was strongly depolarized at the same time. If depolarization was only weak, the synapse showed LTD (Artola et al., 1990). Long-term synaptic plasticity has also been set in relation to pre- and postsynaptic spike timing. If a presynaptic spike preceded repetitive postsynaptic spiking by an amount of 20 ms or less, the synapse under consideration showed LTP. Conversely, if a postsynaptic spike preceded repetitive presynaptic firing by 20 ms or less, LTD resulted. Outside this narrow time window, no change was observed (Markram et al., 1997; Bi and Poo, 1998). In other words, LTP occurred if the presynaptic spikes had a chance to contribute to the generation of the postsynaptic spike, an acausal sequence resulted in LTD. These mechanisms were also shown to depend on NMDA-receptor activity.

Finally, activity-dependent long-term synaptic plasticity was found to be not always strictly confined to the synapses at which activity was paired. Recent studies in rat hippocampus revealed that synapses which undergo Hebbian plasticity, cause neighboring synapses to show a similar modification. This behavior was found for synapses that were adjacent at the same axonal terminal (Bonhoeffer et al., 1989) as well as for adjacent synapses on the same postsynaptic dendritic tree (Engert and Bonhoeffer, 1997). Probably, geometrical rather than electrotonic neighborhood represents the criterion for non-Hebbian synaptic plasticity, and nitric oxide (Schuman and Madison, 1994) is discussed as a possible messenger for its initiation.

In summary, we have seen that neuronal signals undergo a variety of complex and nonlinear transformations along their way through the neuron. Although neurons stimulated with a stationary current show a relatively simple thresholded linear behavior, realistically stimulated neurons which are embedded in a network carry out much more complex operations. A response to the question how these operations can result as an emergent property of the network, can bring us towards a better understanding of brain function. But before we can setup computational neural network models, we must learn more about the biological system which we want to model. We chose the visual system of mammals, and provide a brief introduction to the neurobiology of early vision in the next chapter.

# Chapter 3

# THE EARLY VISUAL SYSTEM OF MACAQUE MONKEYS

The cerebral cortex has the shape of a $2 - 3$ mm thick lamina. It consists mainly of cell bodies and local axonal connections, and shows a relatively uniform structure, as one proceeds laterally across its different areas. Despite its uniformity, different parts of the cerebral cortex perform a large variety of different signal processing tasks (cf. figure 1.1). Hence, it seems desirable to obtain a detailed understanding of one of these parts, because this will probably give strong hints towards the operation principles of other cortical sections.

There are at least two reasons to chose the visual system for investigation. Firstly, vision represents the most important sensory modality of humans, and it is therefore particularly interesting to learn more about our visual system. More importantly, however, the visual system is a relatively well-investigated structure of the brains of higher mammals such as cats and macaque monkeys. Figure 3.1 illustrates the structure of the cerebral cortex of a macaque monkey (top left: side view), which surrounds deeper brain centers as a strongly folded sheet. The visual cortex covers most of the occipital lobe (left part) and can be divided into different visual areas, denoted by V1 - V5, etc. Each visual area forms a complete representation of the whole visual field, though usually different features are represented. Figure 3.1c displays an unfolded and flattened macaque cortex, on which again some visual areas are marked. The visual cortex of macaque monkeys covers roughly 50 % of their entire cortex, a much higher percentage than found in humans. Actually, the visual system of macaques is relatively similar to its human counterpart, and can serve as a good model system for human vision. The leftmost structure in figure 3.1c is the primary visual cortex or V1. It is the largest visual area and represents the first stage of cortical visual processing. We will choose the primary visual cortex of macaque monkeys as a model system for visual information processing. Where data from macaque V1 is not available, we will summarize measurements form cat cortex.

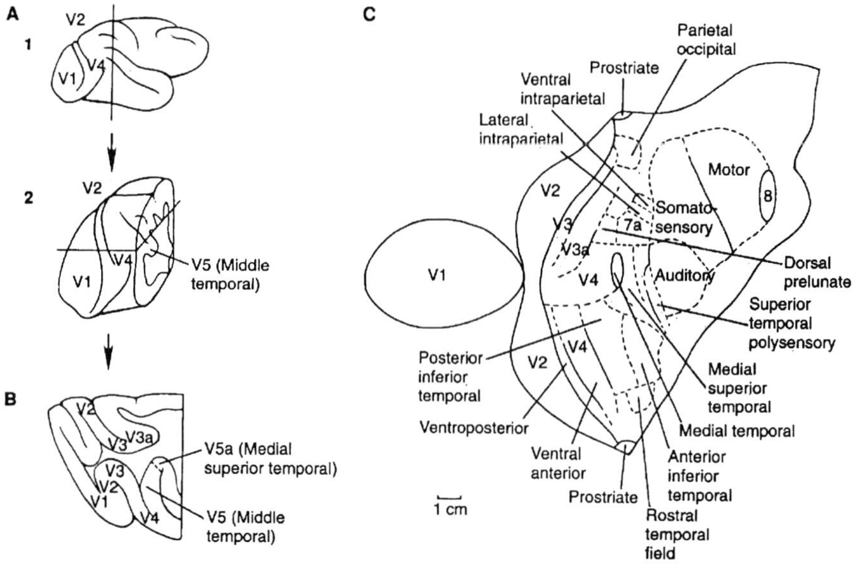


*Figure 3.1.* The visual system of a macaque monkey (after (Kandel et al., 1991)). V1 - V5 denote different visual areas. **(a)** Lateral view of a macaque brain (1) and an expanded view of the occipital lobe (2) (left: posterior; right: anterior). The visual system covers most of the occipital lobe. **(b)** Horizontal section through the occipital lobe and locations of some known visual areas. **(c)** Unfolded and flattened cerebral cortex of a macaque. The visual cortex covers roughly half of its area, and the primary visual cortex, V1, represents the largest visual area.

## 1. Anatomy of the Early Visual Pathway

We start our trip through the visual system of monkeys by summarizing some important aspects of its anatomy. This will allow us to deduct principles of the wiring patterns between neurons which in turn will serve as input to computational models for early vision. A more detailed description of the early visual pathway of mammals can be found in (Kandel et al., 1991).

Figure 3.2 sketches the overall anatomical structure of the early visual pathway. When a cat or monkey fixates its environment, light that falls into the eye is focussed by the cornea and the lens to form two images on the left and right retinae. The part of the world that contributes to the image formed on the two retinae is called the visual field of the animal. Each retina transforms the incoming light intensity distribution into spike patterns, which are transmitted by the two optic nerves into the central nervous system. Each optic nerve consists of roughly $10^6$ myelinated axons, almost all of which target a structure within the

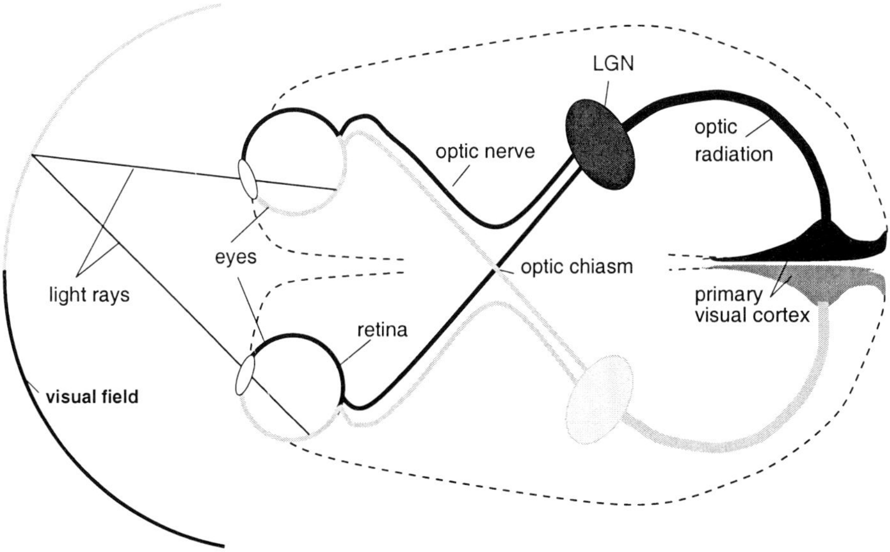


*Figure 3.2.* Schematic sketch of the early visual pathway of higher mammals (adapted from (Bauer, 1999)). For further description see text.

thalamus, which is called the Lateral Geniculate Nucleus or LGN. At the optic chiasm, each optic nerve branches in a way that one half of the fibers targets the "contralateral" LGN at the side opposite to its origin, whereas the rest contacts the "ipsilateral" LGN at the same side as the eye of origin. The cross-over of the nerve fibers occurs in a highly ordered fashion, which ensures that each LGN receives the fibers from the ipsilateral parts of both retinae. Thus, each LGN processes the binocular signals from the contralateral hemisphere of the visual field.

Proceeding from the LGN, another less concentrated bundle of fibers, which is referred to as the optic radiation, contacts the primary visual cortex (also referred to as area 17 or V1) at the occipital lobe of the neocortex. Because these fibers represent input to V1, they are called "afferent fibers" or "afferents". In contrast to the optic nerve, the optic radiation does not cross hemispheres. Hence, each hemisphere of the primary visual cortex processes visual information from the contralateral hemisphere of the visual field. Two major output streams arise from V1. The first stream projects from V1 to higher visual cortical areas, whereas the second stream projects back to the LGN and other subcortical nuclei.

One important feature of the connection pattern between the retina, LGN and V1 is its topographic order or "retinotopy": Fibers that start from neighboring retinal ganglion cells contact neighboring LGN cells which in turn terminate at neighboring regions of V1 (Hubel and Wiesel, 1977). In the following paragraphs, we will consider the anatomy of the retina, the LGN, V1 and the organization of the axonal fibers between them in some more detail.

## 1.1 The Retina

The retina represents the first processing stage of visual signals in the early visual pathway. Besides translating the light intensity distribution into membrane potential changes, the retina carries out important steps of visual processing including adaptation to changes in the light intensity and the detection of color, contrast, and movement of contrast patterns.

Figure 3.3a summarizes the vertical structure of the retina of higher mammals. It consists of three neuronal or granular layers: the photoreceptor layer (top), the bipolar layer with horizontal cells, bipolar cells and amacrine cells (middle), and the retinal ganglion cell layer (bottom). The ganglion cell axons form the optic nerve and transfer the result of retinal processing to the LGN. The three granular layers are separated by plexiform layers, where synaptic connections between the neurons are concentrated. The retina is oriented in a way that incoming light passes all the layers before it finally reaches the photoreceptors (top), where it evokes graded changes in the membrane potential. Photoreceptors can be divided into rods, which are highly sensitive to light, and less light sensitive cones. In many color- sensitive animals such as macaque monkeys, cones have three different spectral sensitivities and provide the basis of color vision.

The signals of the photoreceptors proceed through the retina via two pathways. In the direct or feed-forward pathway, one or several photoreceptors contact a bipolar cell, and one or several bipolar cells in turn drive one ganglion cell. Bipolar cells receive exclusive input either from rods or cones and can be divided into rod- or cone-bipolar cells. Also, the synapses from the photoreceptors to the bipolar cell can be inhibitory or excitatory, which gives rise to a further subdivision of bipolar cells into "ON" and "OFF" bipolar cells. The meaning of "ON" and "OFF" will become more clear in section 2.1. Synapses from bipolar cells to ganglion cells, in turn, are always excitatory: ON (OFF) bipolar cells contact ON (OFF) ganglion cells.

In the indirect or lateral pathway, horizontal cells mediate a lateral information flow between bipolar cells as well as between photoreceptors and bipolar cells. Figure 3.3b provides a simplified scheme of the major wiring patterns together with the signs of the connecting synapses. The indirect pathway acts oppositely to the direct pathway. Amacrine cells serve a multitude of tasks including the selection of the rod and the cone pathway in scotopic (dark-adapted) and photopic (daylight-adapted) vision.

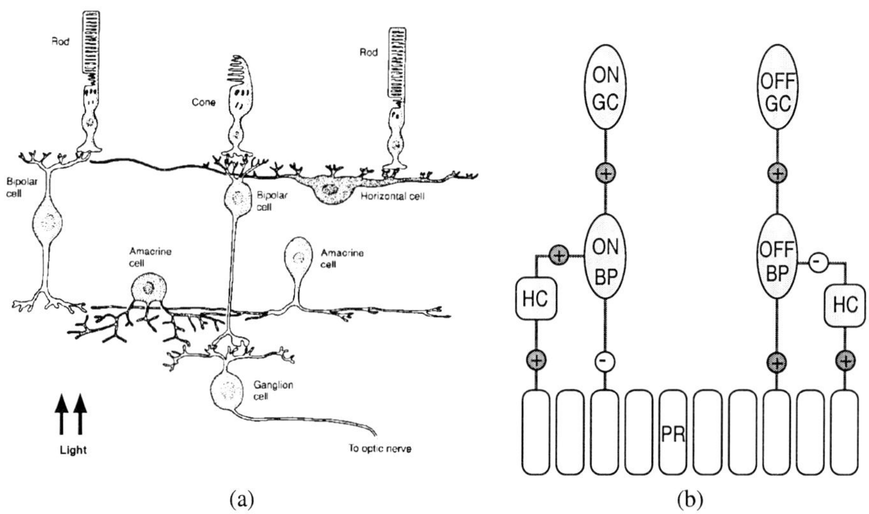


*Figure 3.3.* **(a)** Wiring pattern in the retina of higher mammals: The information flows via the direct pathway from the photoreceptors (top) through the bipolar cells (middle) to the ganglion cells (bottom). The indirect or lateral pathway is mediated by the horizontal and amacrine cells. **(b)** Schematic illustration of the wiring pattern for an ON and an OFF ganglion cell. PR: photoreceptor, HC: horizontal cell, BP: bipolar cell, GC: ganglion cell. circles with "plus"-sign excitatory connection. Circles with "minus"-sign: inhibitory connection.

Besides the subdivision into ON- and OFF- neurons, there exists a third subdivision of retinal ganglion cells into $\alpha$-neurons and $\beta$-neurons, and these cell types show profound functional differences (cf. section 2). Retinal wiring is characterized by a strong convergence: Within each eye, about 120 million photoreceptors, which are densely packed to a next neighbor distance of about 2.5 $\mu$m converge to roughly 1 million ganglion cells.

The lateral organization of the retina is very inhomogeneous: Near the center of the retina, where the lens focuses the image of objects fixated by the eyes, we find the fovea, where visual patterns can be most highly resolved. Many anatomical characteristics change strongly with the distance from the fovea (Wässle et al., 1989), which is measured in angular degrees of eccentricity. Near the fovea, all receptor cells are cones, almost all ganglion cells are $\beta$-cells, and there is the smallest convergence between receptors and ganglion cells. In fact, only one photoreceptor contacts one bipolar cell, which feeds one ganglion cell. With increasing retinal eccentricity, the densities of photoreceptors and ganglion cells decrease, and the fraction of rods and $\alpha$-ganglion cells as well as the convergence rate increases. For example, at the periphery the ganglion cell density is 1000 – 2000 fold lower than close to the fovea, and signals from several hundreds of photoreceptors converge to one ganglion cell.

locations in layer 4 of the primary visual cortex. This retinotopic arrangement is illustrated in figure 3.7a for a human brain, where the top picture subdivides the visual field into 12 regions. The bottom part of the figure shows, in which parts of the primary visual cortex these regions of the visual field are represented. Clearly, the neighborhood relationships are preserved, but the foveal region is expanded compared to the periphery in the cortex. An important measure for the characterization of the retinocortical projection is the magnification factor (figure 3.7b). If two retinal ganglion cells collect information from two patches of the visual field that are 1 deg of visual angle apart from each other, the cortical magnification factor $M$ specifies the average distance of their relayed projections into the cortex. The cortical magnification factor varies strongly with retinal eccentricity.

The details of the cortical wiring patterns strongly depend on the species under consideration. In the next two paragraphs we therefore concentrate on macaque V1 for a slightly more detailed description of the cortical circuitry.

## 1.4 Vertical Circuitry in V1

In macaque monkeys, the six principal layers of the cortex can be further subdivided. Figure 3.8a shows a vertical slice through V1 , which has been stained for the metabolic enzyme cytochrome oxydase (CO). Regions of high CO concentration have been more active on average than others and receive direct thalamic input (Horton, 1984). One observes three horizontal dark bands, which can be assigned to layers 4a, $4c_\alpha$ plus $4c_\beta$, and 6 (see figure 3.8b). Additionally, small CO rich blobs with a diameter of 250 $\mu$m are embedded periodically in the superficial layers (arrows in figure 3.8a).

Figure 3.8b summarizes evidence for excitatory vertical pathways through V1 of monkeys (Blasdel and Lund, 1983; Lund, 1987; Yoshioka et al., 1994; Levitt et al., 1996; Yabuta and Callaway, 1998), for reviews see (Lund, 1988; Callaway, 1998). The afferent fibers from the LGN target five different zones, which are fully segregated for the M, P, and I type LGN fibers (left). Magnocellular fibers terminate in layer $4c_\alpha$ and send collaterals to layer 6. Their individual axonal fields have a diameter between 600 and 2000 $\mu$m and each axon forms about 3000 synapses in total (cf. (Blasdel and Lund, 1983), see also (Stetter et al., 1993) and references therein). There are probably two populations of parvocellular fibers. The first population includes 90 % of the fibers, which spread about 250 $\mu$m in layer $4c_\beta$ (about 300 synapses), the remaining 10 % of the fibers terminate in layer 4a. Intercalated fibers project directly into the CO blobs. Afferent fibers from the left and the right eyes are laterally segregated into alternating stripes of about 450 $\mu$m, which are called ocular dominance (OD) stripes (Hubel and Wiesel, 1968; Hubel and Wiesel, 1977). They are not shown in figure 3.8b. Instead, the illustration can be viewed as a slice parallel to one of the ocular dominance stripes. Ocular dominance segre-

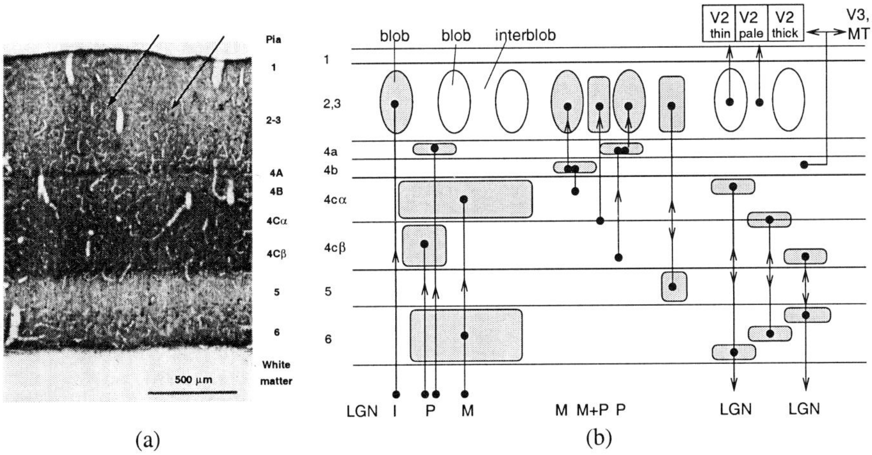


*Figure 3.8.* **(a)** vertical slice through V1 of a macaque (from (Bauer, 1999), after (Blasdel and Lund, 1983)). Darker regions are stained for cytochrome oxidase (CO) and mark regions of thalamic input. **(b)** Cortical layers of macaque V1 together with the major excitatory streams through V1.

gation is most evident physiologically in layer 4c and decreases as projections proceed to the superficial and deeper layers.

Two major excitatory intracortical pathways can be separated: $(i)$ An upward path (figure 3.8b middle): The first stream from layer 4 to the superficial layers includes a redistribution of the three incoming pathways to three new streams. One M-dominated stream projects from upper $4c_\alpha$ to 4b and then predominantly to extrastriate areas, a second stream collects information from mid 4c (M + P) and projects to interblob zones, and the third P-dominated stream projects from lower $4c_\beta$ over 4a to the blobs. $(ii)$ A downward path (figure 3.8b right): The same three zones of layer 4c project to three distinct divisions of layer 6 and receive recurrent feedback-connections from their target layer 6 neurons (Wiser and Callaway, 1996; Callaway, 1998). There is also a large variety of inhibitory interneurons, which project between the laminae (Lund, 1987). Important inhibitory pathways implement inhibitory feedback between layers $4c_\alpha$ and $4c_\beta$ and between layers 6 layer 4. In combination with local lateral connections, interlaminar inhibition might serve as an anatomical substrate for perceptual grouping, as suggested by theoretical studies based on "competitive layer" neural network models (Wersing et al., 1997; Ontrup and Ritter, 1998; Wersing et al., 2000). Output fibers from V1, finally, project from layers 4b, 2, and 3 to areas V2, V3 and MT, and project back to the LGN from layer 6.

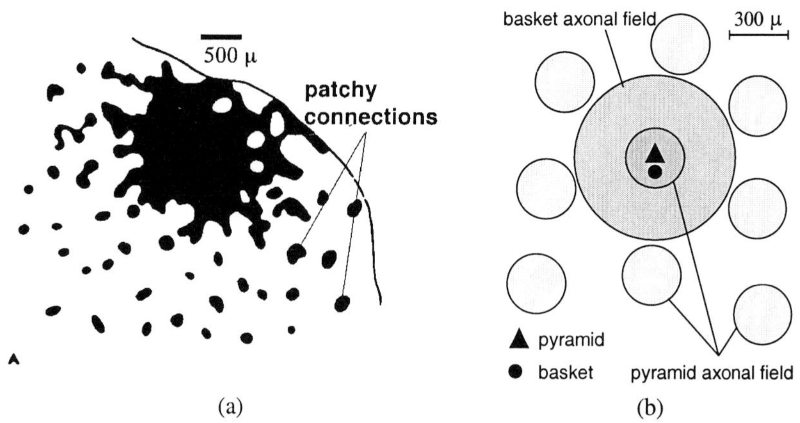


*Figure 3.9.* **(a)** Pattern of lateral connectivity of cortical pyramidal cells (adapted from (Rockland and Lund, 1983)). Black patches represent the reconstruction of HRP- labeled collaterals of pyramidal cells in tangential section. **(b)** Schematic diagram of excitatory and inhibitory lateral connectivity in the superficial layers. The cortex is viewed from top.

## 1.5 Lateral Connectivity in V1

Excitatory lateral intracortical connections are mediated by collateral fibers of pyramidal neurons. Pyramidal neurons in the superficial layers and excitatory stellate cells in upper layer 4 express two sets of collaterals: $(i)$ A dense and local lateral connectivity, which is restricted to a diameter of approximately $300 - 400\ \mu$m. $(ii)$ Long-range lateral fibers, the synaptic boutons of which are restricted to a set of patches with approximately 250 $\mu$m diameter and 250 $\mu$m gaps in between, but which can extend up to 3 mm laterally from each neuron. Because of the patchy appearance of the clusters of boutons, these long-range connections are referred to as patchy connections. Figure 3.9a displays HRP-labeled patchy connections of a set of stained pyramidal cells. Long-range connections change from a patchy shape in the superficial layers to a more elongated shape in layer 4b. Long-range connections in layers 5 and 6 lack the strongly patterned connectivity seen in the superficial layers.

Lateral inhibition is present in upper layer 4 and the superficial layers. It is mediated by basket neurons with an axonal field of 600 – 800 $\mu$m in diameter (Lund, 1987). Local inhibition is mediated by other neuron types including chandelier cells, which spread their axons only by about 200 $\mu$m. V1 contains a large variety of inhibitory interneurons, which are reviewed in (Lund, 1987). Figure 3.9b summarizes the lateral connection scheme in the superficial layers. Lateral inhibition spreads somewhat wider than the diameter of the intracortical patches driven by pyramidal neurons and may help create the patchy pattern of connections (Lund et al., 1994). Patchy connections probably contribute both to local visual processing within the classical receptive field (cf section 2.4) and to the processing of visual context (Lund et al., 1995; Bartsch et al., 1997; Stetter et al., 2000c).

## 2. Physiology of the Early Visual Pathway

After summarizing the wiring pattern of the early visual pathway, we will now consider the response properties of neurons at different levels of early vision. The combination of knowledge about both the connectivity and the neuronal responses that are generated by this connectivity in the presence of a stimulus will serve as important inputs for the design of mathematical models of early vision.

### 2.1 Retinal and Geniculate Neurons

If retinal ganglion cells are successfully excited by a visual pattern, they fire a sequence of action potentials or spikes, which is called a spike train. A spike train of a neuron can be detected by putting a microelectrode close to its cell body or axon and measuring the extracellular electrical potential over time. In cats each retinal ganglion cell is sensitive to stimuli only within a small, roughly circular part of the visual field, which is called its receptive field (Kuffler, 1953). Also, if this receptive field was illuminated with a small bright spot, the response was not uniform but depended on its position within the receptive field. Figure 3.10 summarizes the change in the neuronal response of ganglion cells to different stimulus conditions. Some neurons were excited, when the spot hit the center of the receptive field, whereas they were silenced by illumination of the surround. These ganglion cells are called ON-center ganglion cells. A second type, referred to as OFF-center ganglion cells, was inhibited by illumination in the center and excited by surround illumination. Diffuse illumination evoked almost no response. A comparison of this center-opponent organization with the anatomical wiring suggests that the receptive field center is mediated by the direct path, whereas the surround is mediated by the lateral path.

The description of the receptive field must include at least two parts: $(i)$ its spatial extent, and $(ii)$ information about the structure of the neural response in space and time. The receptive field of a retinal ganglion cell is small and approximately concentric. For ON-ganglion cells it is excitatory in the center and inhibitory in the surround. This arrangement is called a "Mexican hat" profile.

The response properties of retinal ganglion cells can be further subdivided according to additional characteristics (DeMonasterio and Gouras, 1975; Croner and Kaplan, 1995), for a review of early work see (Kaneko, 1979). About 80 % of the cells have comparatively small receptive fields, a high luminance contrast

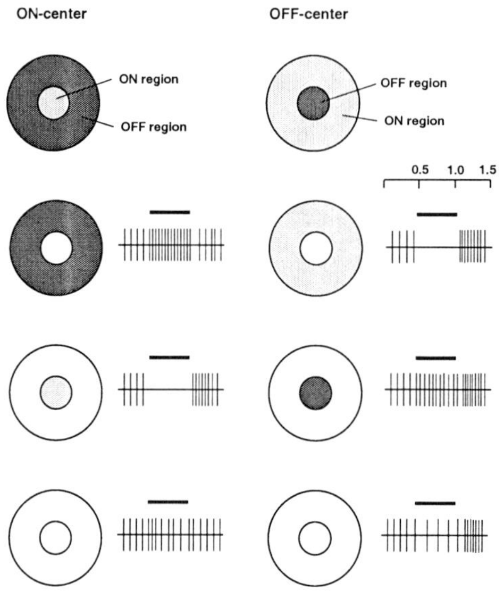


*Figure 3.10* Receptive field organization of retinal ganglion cells (from (Bauer, 1999) after (Kuffler, 1953)). Left column: ON-center cells are excited by illumination in the center (top), inhibited by illumination of the surround (middle), and do not respond strongly to diffuse illumination (bottom). Right column: OFF-center cells show the opposite characteristics.

threshold, respond linearly to stimuli, and are color-sensitive. For example, in one class of these cells, the center is excited best by red light, and the surround is maximally inhibited by green light. This class, which is assigned the abbreviation "r+g-" cells, will respond both to a contrast in white light illumination or to a diffuse red illumination. Other classes of color-sensitive ganglion cells are cells with g+r-, y+b-, and b+y- characteristics, where b and y denote blue and yellow. The same four characteristics can be found for OFF-center neurons. It was found, that these neurons correspond to the $\beta$-cells, which had been identified anatomically. Approximately 10 % of the retinal neurons had large receptive fields (3-5 times as large as $\beta$-cells at the same eccentricity (DeMonasterio and Gouras, 1975)), a higher sensitivity to luminance contrast, a fast response to stimulus changes, and no color-sensitivity. Some of these neurons respond only to the onset or the termination of the stimulus. Another class responds nonlinearly both to the onset and to the termination. These transient neurons have been identified as $\alpha$-cells. The remaining 10 % of neurons do not project to the LGN and will not be further considered here.

The two different receptive field properties of $\alpha$- and $\beta$-neurons suggest that they are devoted to different tasks of visual information processing. $\alpha$-ganglion cells are designed to sensitively detect motion, even of structures with low contrast. They act as novelty detectors, which report quickly about any changes in the environment, and can be viewed as an alert system. $\beta$-ganglion cells, in contrast, are optimized for a detailed analysis of a more or less stationary scene. Consequently, the distribution of both cell types strongly depends on the retinal eccentricity: The fovea contains only $\beta$-cells, which help analyzing

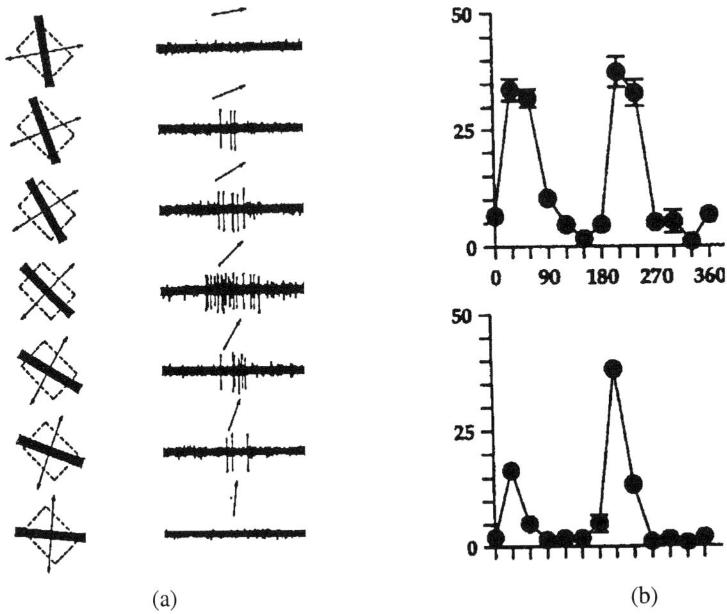


*Figure 3.11.* **(a)** Response of a complex orientation selective cell of macaque V1 (from (Hubel and Wiesel, 1968)). The black bar illustrates the stimulus, the rectangle marks the receptive field borders. The neuron is strongly tuned to the orientation of the bar. **(b)** Orientation tuning curves of two V1 cells (from (Levitt and Lund, 1997)). x-axis: stimulus-orientation in degrees (90 deg refers to a horizontal grating moving upwards). y-axis: Spike rate (1/s). Both neurons are strongly tuned to orientations near 30 deg. In addition, the bottom neuron is direction selective.

the fixated object in some detail, whereas the fraction of $\alpha$-cells increases with eccentricity (DeMonasterio and Gouras, 1975). Consistent with the amount of convergence to ganglion cells, the receptive field sizes of both $\alpha$ and $\beta$ cells strongly increase with eccentricity. The center size of $\alpha$ receptive fields ranges from about $0.12 - 2$ deg of visual angle for eccentricities between 1 and 40 deg. The smallest $\beta$ receptive field centers in the fovea are less than one minute of arc in diameter.

In monkeys, the receptive field profiles of M and P neurons of the LGN are very similar to those of the retinal $\alpha$ and $\beta$ neurons, respectively (Kaplan and Shapley, 1982). We will not summarize them separately here, but it shall be mentioned that the responses of LGN neurons, in contrast to retinal neurons, are due to substantial modulatory influences from the cortex, the TRN, and other brain structures (cf section 1.2), which could participate in the control of attentional states.

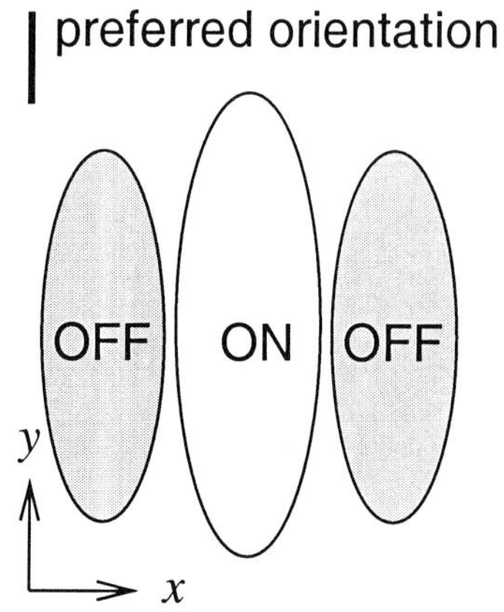


*Figure 3.12* Sketch of a simple receptive field profile. Typically, simple cells have two or three elongated and segregated subfields, which show alternatingly ON- and OFF response. The subfield can be less regular than in the sketch.

## 2.2 Cortical Receptive Fields

Similar to the retinal and the LGN-cells, neurons in the primary visual cortex respond only to stimuli within a small patch of the visual field, which is called their "classical receptive field". Also, most of them can be driven from either eye although one eye usually dominates the input. But compared to LGN neurons, cortical response characteristics show a qualitatively new feature: most of the cortical neurons respond selectively to contrast lines, bars, or gratings of a certain orientation within the receptive field (Hubel and Wiesel, 1962; Hubel and Wiesel, 1968; Dow, 1974; Schiller et al., 1976a; Schiller et al., 1976b; Hubel and Wiesel, 1977). Depending on the subtype of neuron, the response to circular spots is only small and diffuse illumination does not drive the neurons at all. Besides orientation-selective neurons, the primary visual cortex contains also some weakly tuned or untuned neurons. Most of these neurons are input neurons in layer $4c_\beta$ and color-opponent cells (similar to LGN neurons) in the superficial layers.

Figure 3.11a shows the spike trains of a cortical neuron in macaque V1 in response to a moving oriented bar. The neuron responds most strongly for a given orientation of the bar, and its response declines strongly with increasing differences of the stimulus orientation from the preferred orientation of the cell: the cell is tuned to the stimulus orientation. In figure 3.11b the average spike rates of two different cortical neurons are plotted as a function of the orientation of a moving grating as stimulus. These curves are called the "orientation tuning curves" of the cells. The response of the top neuron is tuned to an orientation of approximately 30 deg. The neuron responds similarly to an orientation of 210 deg, which corresponds to a grating of the same orientation than the 30 deg stimulus but moving into the opposite direction. The bottom plot shows the response of a neuron that responds differently for the two opposite directions of motion of an optimally oriented grating – it is direction selective.

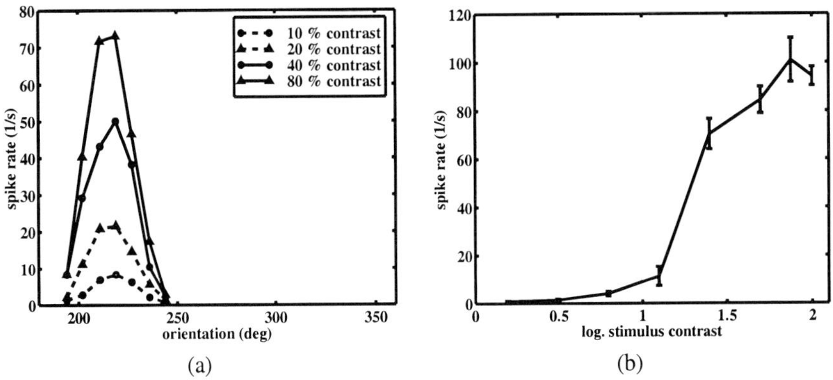


*Figure 3.13.* **(a)** Orientation tuning curve of a cell in the primary visual cortex of a cat, measured for different levels of stimulus contrast (from (Sclar and Freeman, 1982)). The tuning width of approximately 45 deg is contrast invariant. **(b)** Contrast-response curve for a neuron from macaque primary visual cortex (Levitt, unpublished result). While LGN neurons respond logarithmically to contrast changes, most of the cortical neurons show a stronger than logarithmic saturation.

The receptive fields of orientation selective neurons can be subdivided into two types: $(i)$ Simple receptive fields consist of two or three elongated and segregated subfields with ON and OFF-characteristics (Hubel and Wiesel, 1962; Hubel and Wiesel, 1968; Schiller et al., 1976a; Jones and Palmer, 1987a; Jones et al., 1987; Jones and Palmer, 1987b). They are sensitive to the position of the stimulus within the receptive field, or, in case of a grating, to the phase of the grating. $(ii)$ Complex cells have larger receptive fields, which cannot be subdivided into ON and OFF-regions. They are insensitive to the stimulus phase, are more likely to be direction selective and respond strongly only to moving stimuli. For both types of neurons, the full width at half maximum of the orientation tuning curves in monkeys varies between 10 and 100 deg, the most frequent tuning strength being 40 deg (Schiller et al., 1976b). However, a considerable number of neurons are tuned as tightly as 20 deg. Similar numbers for the strength of orientation tuning hold for cats (Hammond and Andrews, 1978).

In cats, orientation tuning is independent of the stimulus contrast (Sclar and Freeman, 1982; Skottun et al., 1987). This is demonstrated in figure 3.13a, which shows the orientation tuning curve of a neuron from the primary visual cortex of a cat. Also, the responses of orientation selective neurons both in cats and monkeys were found to saturate strongly with increasing stimulus contrast (Albrecht and Hamilton, 1982; Albrecht, 1995). This saturation of the cortical contrast-response function is usually stronger than that of LGN cells, the contrast-response function of which depends linearly on the log contrast. Figure 3.13b shows the mean response of a neuron in macaque V1 as a function of log contrast. The saturation at high contrast levels is stronger than logarithmic, and for very high contrast levels, the response even decreases in many cases. This behavior, which can be seen at the right part of figure 3.13b is referred to as supersaturation. The maximum spike rates at 100 % contrast remain well below the physiologically possible limit. This allows to hypothesize that contrast saturation is not caused by the properties of single neurons but rather by the properties of the whole neural network.

The working point of the contrast-response curve (the site of its steepest slope) can change depending on the history of a neuron. If a receptive field in cat area 17 is exposed to high contrast stimulation for a longer time, the contrast response curve shifts to higher contrast levels and the neuron becomes less contrast sensitive (Ohzawa et al., 1985; Sengpiel et al., 1998a).

## 2.3 Cortical Response to More Complex Stimuli

When stimuli with a more complex structure than a simple oriented grating are presented, cortical neurons show interesting changes of their response. These properties can provide the basis for a deeper understanding of how visual signals are integrated in the recurrent cortical network.

Figure 3.14 summarizes results by (DeAngelis et al., 1992), who presented a linear superposition of two grating stimuli with orthogonal orientations to the classical receptive field of a neuron in cat primary visual cortex. The plots in figure 3.14a show, how the response of the neuron increases with increasing length (top) or width (bottom) of a single, optimally oriented grating. The stimulus size, at which the responses saturate, define the size of the classical receptive field. The plots in figure 3.14b show that the response of the neuron is strongly suppressed, if a second grating orthogonal grating is superimposed on the first grating. This effect will be henceforth referred to as cross-stimulation suppression.

Stimuli that lie outside the classical receptive field of a neuron usually are unable to evoke a response of this cell. However, a number of studies on monkeys (Kapadia et al., 1995; Sillito et al., 1995; Levitt and Lund, 1997) and cats (Blakemore and Tobin, 1972; Gilbert and Wiesel, 1990; Polat et al., 1998) have shown, that the presence of such a flanking stimulus can considerably modulate the response of a cell to a stimulus within its classical receptive field. In other words, the response of the cell to a local feature depends on the visual context into which this feature is embedded. Therefore, this class of phenomena is often referred to as contextual effects.

If the centered stimulus is surrounded by an annular high-contrast stimulus outside the classical receptive field, and if the surround orientation matches the center orientation, the response is suppressed. This effect is called iso-orientation suppression. For orthogonal orientation, the response is slightly (Levitt and Lund, 1997) or strongly (Sillito et al., 1995) facilitated. This effect

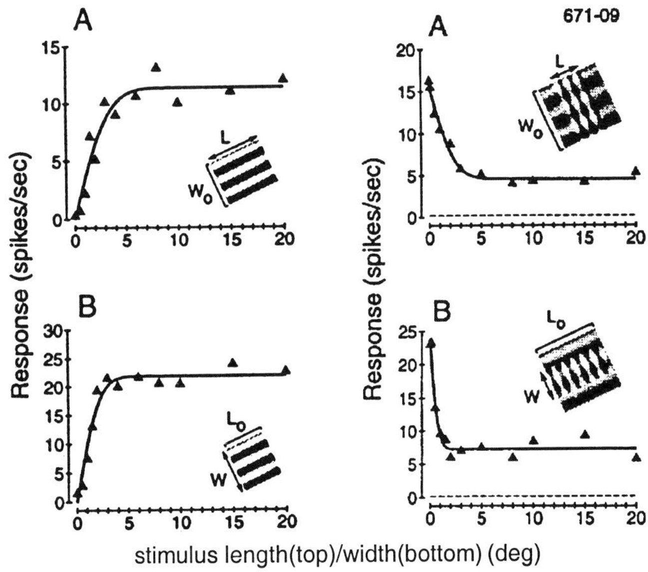


(a) (b)

*Figure 3.14.* **(a)** Response of a cortical neuron to a small and optimally oriented grating as a function of the stimulus length (A) and width (B). **(b)** Neuronal response, if an orthogonal grating is superimposed on the original oriented grating. Along the abscissa, the length (A) or width (B) of the orthogonal grating is varied. The response is strongly suppressed by the presence of the second grating (from (DeAngelis et al., 1992)).

is called cross-orientation facilitation. The black filled circles in the diagrams Figure 3.15a demonstrate these effects: Two neurons have been stimulated with a high contrast center stimulus and an annular non-classical surround. The plots show their responses as a function of the surround orientation. If the center and surround orientations coincide, the responses of the cells decrease compared to presentation of the center alone (solid line), whereas for orthogonal orientation the response is unchanged or slightly facilitated. The open circles are the responses to presentation of the surround alone. They demonstrate that the surround has only a modulatory effect and cannot drive the neurons alone. Figure 3.15b summarizes a related effect which is called end-stopping. The plots show, how the neuronal response varies with the length (E) and width (F) of an optimally oriented stimulus. The response reaches its maximum, when the stimulus size fits the dimensions of the classical receptive field. When the stimulus further increases in length or width, it begins to cover part of the non-classical surround of the cell and causes a suppressive effect. Not all cells show end stopping. For example, the neurons in figure 3.14a showed no contextual

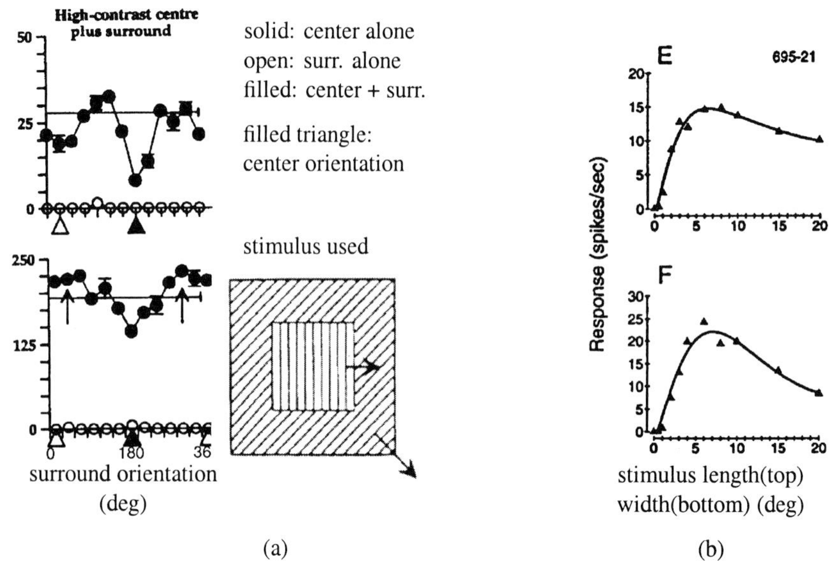


*Figure 3.15.* **(a)** Response of two cells from macaque V1 to a center stimulus as a function of the orientation of an annular surround outside the classical receptive field (black filled circles; stimulus geometry is shown at the bottom right; from (Levitt and Lund, 1997)). Solid line: Response to stimulation by the center alone. Open circles: Response to stimulation by the surround alone. Black filled triangles mark the fixed center orientation. Both cells show iso-orientation suppression, and a slight facilitation for orthogonal orientations. Stimuli had high contrast (75 %); Differences of 180 deg in orientation refer to the same orientation but opposite directions of movement. **(b)** End stopping in cat visual cortex (from (DeAngelis et al., 1992)). The plot shows the response of neurons as a function of the length (E) and the width (F) of an optimally oriented grating. Beyond an optimal length or width, the stimulus contributes to the nonclassical surround and causes a suppression.

modulation, and other neurons have been reported to be strongly facilitated by co-aligned flanking bar stimuli (Kapadia et al., 1999).

If the center contrast is low, contextual effects can change their characteristics. Some studies in cats (Polat et al., 1998) and monkeys (Kapadia et al., 1995) report, that iso-orientation suppression turns into facilitation for low center contrast, which is consistent with a fill-in-paradigm. Other studies (Levitt and Lund, 1997) report cross-orientation facilitation to turn into suppression.

Contextual effects depend also on the geometrical properties of the stimulus in the non-classical surround. If the center stimulus is accompanied by two small flanking bars co-aligned with the central oriented stimulus, iso-orientation facilitation has been observed (Polat et al., 1998; Kapadia et al., 1999). In chapter 10, we will address the possible role of iso-orientation modulation for object-based image segmentation.

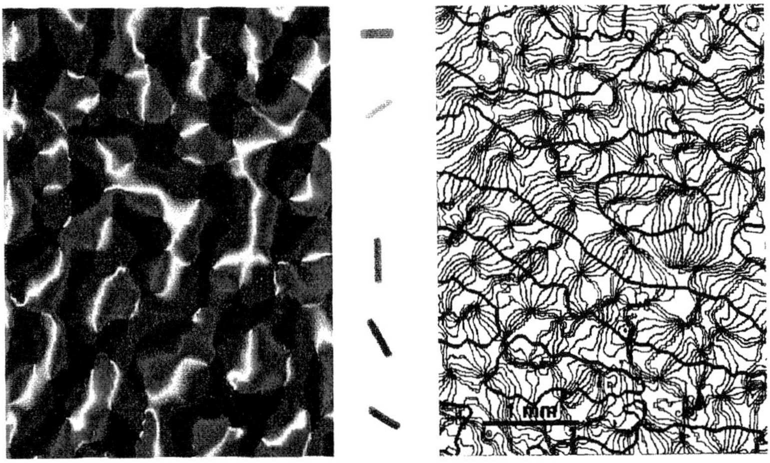

(a) (b)

*Figure 3.16.* **(a)** Orientation map of macaque V1. Colors code different preferred orientations as indicated by the colored bars. The horizontal extent is 3.3 mm **(b)** Superposition of ocular dominance and orientation maps. Thick lines: borders of ocular dominance stripes. Thin lines: Iso-orientation contours within the orientation map of the same animal (from (Obermayer and Blasdel, 1993)). *(For color figure see Color Section,* p. xv*)*

## 2.4 Spatial Distribution of Cortical Response Properties

The distribution of the receptive field properties across the cortex was first investigated by electrode penetrations (Hubel and Wiesel, 1962; Hubel and Wiesel, 1968). In this technique a microelectrode is advanced into the tissue in steps of a few micrometers and the receptive field properties of many neurons are recorded sequentially. The actual trajectory of the microelectrode is reconstructed after the measurement from brain sections by analyzing small lesions which had been generated at the site of the electrode tip during the measurement.

These penetrations revealed a columnar functional structure of the cortex. Neurons that lie within a narrow column of about 20-30 $\mu$m orthogonal to the cortical surface, often show similar preferred orientations and are driven preferentially by the same eye. Each such column is called an orientation column. However, the columnar structure is not strictly preserved. i.e., orientation selectivity sometimes changes as one proceeds vertically along the column (Bauer et al., 1989). Also, the complexity of the neuronal orientation-dependent response and some other feature selectivities change within the column, i.e. they are different for the different cortical layers. In monkeys, most of the layer 4a and 4c$_\beta$ neurons, and some of the 4c$_\alpha$ neurons are spatial opponent rather than orientation selective (Blasdel and Fitzpatrick, 1984). Simple and complex

Chapter 4

# OPTICAL IMAGING OF BRAIN ACTIVITY

The last chapter has made clear, that the visual cortex is a highly structured network of many millions of neurons. In order to understand the function of this complex system, it is very important not only to know responses of individual neurons to stimulation but also to be able to measure how the visual system responds to given stimuli as a whole. This reflects the hypothesis that the visual cortex represents many aspects of visual information by the activity of neuron populations rather than individual neurons (cf. chapter 3, section 1.3).

During the last two decades, optical imaging of neural activity (Blasdel and Salama, 1986; Grinvald et al., 1986) has been proven a powerful technique for the detection of two-dimensional neural activation patterns with sub-millimeter resolution. Being the only in vivo method for functional imaging which can spatially resolve the details of the functional cortical architecture (as opposed to fMRI and PET (Cohen and Bookheimer, 1994; Ostergaard et al., 1998), but see (Kim et al., 2000a) for a promising fMRI study and its controversial discussion (Logothetis, 2000; Kim et al., 2000b), optical imaging has triggered the discovery of many important aspects of the functional organization of the cortex such as the two-dimensional pinwheel-like structure of orientation maps in cats (Bonhoeffer and Grinvald, 1991; Bonhoeffer and Grinvald, 1993; Bonhoeffer et al., 1995; Kim and Bonhoeffer, 1994), monkeys (Blasdel and Salama, 1986; Blasdel, 1992a; Blasdel, 1992b), ferrets (Chapman et al., 1996; Rao et al., 1997) and other species (Bosking et al., 1997), the layout of direction selectivity maps (Shmuel and Grinvald, 1996; Weliky et al., 1996) and the relationship between different map systems within one and the same animal (Yoshioka et al., 1994; Yoshioka et al., 1996; Obermayer and Blasdel, 1993; Hübener et al., 1997). Optical imaging has been successfully applied to reveal the structure of auditory (Hess and Scheich, 1996; Dinse et al., 1997) and somatosensory maps (Godde et al., 1996) and helped to gain useful insights into developmental processes of cortical function (Kim and Bonhoeffer, 1994; Gödecke and Bonhoeffer, 1996; Gödecke et al., 1997; Dinse et al., 1997; Godde et al., 1996). Finally, the technique allowed for the first time to directly relate anatomical cortical wiring patterns to functional aspects of the cortex (Malach et al., 1993; Malach et al., 1994; Kisvarday et al., 1994; Das and Gilbert, 1995; Dalva et al., 1997).

Therefore, optical recording of brain activity and accompanying theoretical methods will be treated in some detail in the present and the four following chapters. Here, I will provide a brief overview over a typical experimental setup and over different biophysical mechanisms that have been exploited for the optical detection of brain activity. I will then present a mathematical framework based on tissue optics, which allows us to quantify the relationship between these biophysical signal sources, their spatial location, and the actual optical measurements. We will see that optical imaging can only access activity patterns near the brain surface, and will use tissue optical considerations to suggest an improved technique for optical imaging of neural activity, which can be used to measure the distribution of neural activity over depth.

## 1. Principles of Optical Imaging

Generally, techniques for the optical recording of brain activity use a relationship between neuronal activity within brain tissue and changes in the optical properties (e.g. the light reflectance) of the brain tissue for light in the visible or near-infrared range of wavelengths. These changes can be evoked by chemical agents applied to the tissue or by metabolically evoked variations in the biological and chemical structure of the tissue itself. Optical imaging aims at the measurement of two-dimensional distributions of activity across large neuron populations and has to be distinguished from the optical recording of activity at the single cell level (Cohen et al., 1968; Cohen, 1973; Fromherz and Vetter, 1992; Fromherz and Müller, 1994). We make this distinction explicit by abbreviating the optical recording of population activity as "optical imaging".

### 1.1 Experimental Setup and Data Collection

Figure 4.1 illustrates a typical experimental setup used for optical imaging. The part of the brain brain surface under consideration is homogeneously illuminated by a narrow-band light source. Because the changes in reflectance we want to detect can be very small, the power of the light source has to be stabilized: for intrinsic signals, variations in the light intensity should not exceed 0.01 % of the mean intensity. Also, because we want to measure changes that occur within the tissue, the light should have a sufficient penetration depth into the tissue. This restricts the choice of the light wavelength to ranges outside the absorption bands of water, between 500 nm and 1300 nm.

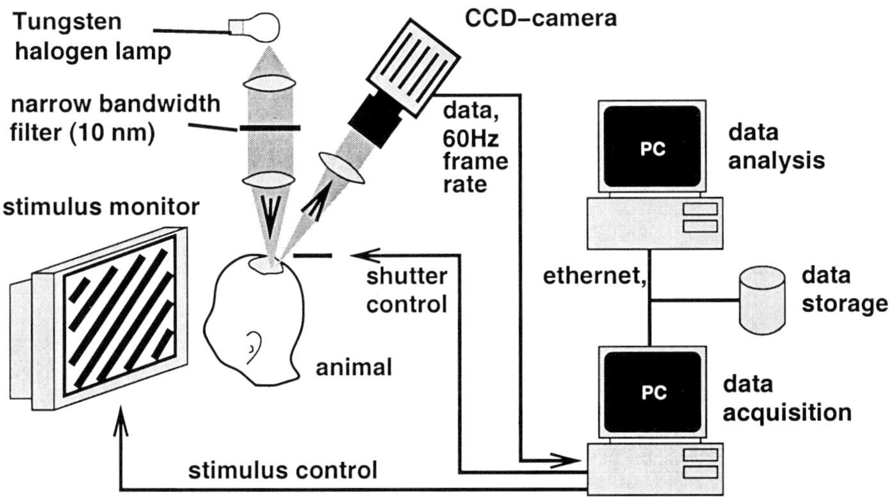


*Figure 4.1.* Typical setup for optical imaging of neuronal activity: Brain tissue is homogeneously illuminated by monochromatic red or near infrared light. A sensitive CCD-camera repeatedly records the light reflected from the brain surface, while the corresponding nerve tissue is stimulated (example shown: visual stimulation). The collected sequences of images are stored and processed on a computer system.

For a single trial of an optical imaging experiment, the brain tissue is stimulated, while a sensitive CCD-camera collects a sequence of images of the tissue before, during and after stimulation, which are stored in a computer system as an image stack. Figure 4.1 shows the example of optical imaging of the visual cortex and its visual stimulation. Stimulation, data collection and data storage are synchronized by the data acquisition computer, which may serve additional control tasks including breath control and synchronization of the anesthetized animal during the trial. After a recovery period, the next trial can be carried out.

The changes in reflectance caused by active neuron clusters can be as small as 0.1 % of the reflected light. Therefore, some care must be taken for the choice of the correct camera properties and the design of an appropriate measurement strategy. Optical signals result from local changes in light absorption, light scattering or fluorescence and their strengths scale linearly with the total amount of light reflected back from the tissue. Hence, in order to maximize the amplitudes of these signals, optical imaging has to be carried out under strong illumination conditions. In this regime, the dominating source of noise for the detection of the light intensity is photon shot noise (Bonhoeffer and Grinvald, 1996; Stetter and Obermayer, 1999). We now have to ask, which statistical error is introduced by photon shot noise and how we can minimize it. If the photons are remitted independently from the constantly illuminated tissue, the number $n$ of photons collected per unit time by a pixel of the CCD camera follows a Poisson distribution. If the expectation of the number of measured photons is $N$, the standard deviation of $n$ around this value is given by $\sigma_n = \sqrt{N}$ (Papoulis, 1965). Hence, the relative statistical error for the measurement of the light reflectance is given by $\sigma_n/N = N^{-1/2}$. In order to measure changes in reflectance with an accuracy of 0.1 %, we have to collect at least $10^6$ photons per pixel and image, and for an accuracy of 0.01 % we need to collect $10^8$ photons.

Due to this reason, the camera should be able to collect as many photons per pixel as possible. This can be achieved by two means or a combination of both: $(i)$ The camera has a high well-capacity $c_w$, which is the maximum number of optically induced electrons that can be stored by a pixel before readout. Usually the number of levels for digitization reflect the signal-to-noise level that is reachable by the camera based on its well-capacity. For example, a 12-bit camera brings the shot noise error down to approximately $1/2^{12} = 1/4096$ under maximal illumination. $(ii)$ The camera should have a high frame rate. If the camera can record $M$ frames per second, but we want to store only $K$ images per second for the image stack, we can obtain an effective well-capacity of $c_w M/K$ by summing up $M/K$ subsequent frames for each image. For both cases, the illumination power should be adjusted to a level which fills the pixels of the camera almost to its well-capacity.

## 1.2 Optical Imaging using Dyes

One class of methods for optical imaging actively generates a relationship between neuronal activity and the optical properties of nerve tissue. This is achieved by probing various biophysical and physical parameters of nerve cells by dyes. Depending on a cellular parameter, these dyes change their spectral absorption, the strength of fluorescence, the wavelength of fluorescence or other optical properties, which can be detected by optical imaging. If the cellular parameter is related to neuronal activity in a well-defined way, the dye can be used to optically probe neuronal activity. Two important representants for dye-imaging are methods based on voltage-sensitive dyes and $Ca^{2+}$-sensitive dyes.

Voltage-sensitive or potentiometric dyes consist of molecules which attach themselves to the cell membrane and change their optical properties depending on the electrical field in their neighborhood (Stark-effect) (Salzberg et al., 1973). If the nerve cell is activated, its membrane potential changes and the electrical field near the membrane is altered. This local field is probed by molecules of the voltage-sensitive dye, which changes its electronic state and therefore its absorption or fluorescence properties. The same behavior can be observed, if a whole stained neuron population is activated, and the recorded signal can be used for dye imaging.

One important dye used for optical imaging include merocyanine oxazolone (NK2367) (Blasdel and Salama, 1986; Blasdel, 1992a). Voltage sensitive dye imaging directly monitors the electrical state of the neuron and provides a high temporal resolution, because the dyes molecules change their state very fast. However, voltage sensitive dye imaging also has some disadvantages: First, the signals do not only reflect the action potential activity, but also monitor all fluctuations in the dendritic membrane potentials (cf. chapter 2, section 3), which may be sub-threshold and possibly do not contribute to the neuronal spiking activity. Also, voltage-sensitive dyes are phototoxic which restricts their applicability to long and chronical measurements. Finally, pure dye signals are hard to detect, because the reflectance signal is superimposed with intrinsic optical signals (cf. chapter 2, section 1.3).

Another version of dye-imaging uses $Ca^{2+}$-sensitive dyes (Tsien, 1980; Poenie, 1992; Galizia et al., 1997) and is referred to as $Ca^{2+}$ imaging. It uses the facts that the postsynaptic intracellular $Ca^{2+}$-concentration increases when the cells are activated, and that a presynaptic calcium-influx is necessary for transmitter release. The activity-dependent increase of intracellular $Ca^{2+}$-concentration both in neurons and glial cells can be due to several different mechanisms including: $(i)$ The influx of free calcium ions from the extracellular space or the endoplasmatic reticulum with relatively short time constants usually in the sub-second range (Kandel et al., 1991; Bygrave and Benedetti, 1996); $(ii)$ The release of $Ca^{2+}$-ions from various internal stores, which can have very slow time-constants in the range of several seconds (Friel and Tsien, 1992). For $Ca^{2+}$ imaging, neural population activity is recorded by introducing a $Ca^{2+}$-sensitive fluorescent dye to the intracellular space of neurons. Evoked changes in the concentration of intracellular free calcium cause a change in the intensity or a spectral shift of the fluorescent light, which can be detected by a CCD-camera and a spectral filter. Calcium-imaging has the advantage of providing much stronger optical signals (a few percent of the background light) than voltage-sensitive dye signals or intrinsic signals (0.1 percent). However, $Ca^{2+}$-imaging detects only an indirect measure of neuronal activity. This measure may differ in its spatial pattern from the actual pattern of activity and may be nonlinearly and even non-uniquely related with the strength of neural activity. Further, the dyes bind $Ca^{2+}$ ions and thus interfere with the cellular machinery to some degree. Finally, $Ca^{2+}$-sensitive dyes are due to both reversible and irreversible photobleaching (Pawley, 1995). Photobleaching imposes constraints on the strength and duration of illumination, limits the signal-to-noise ratio for photon shot noise and interferes with the detection of activity-related changes in fluorescence. In chapter 8, we will show, how these problems can be overcome by use of statistical data analysis techniques.

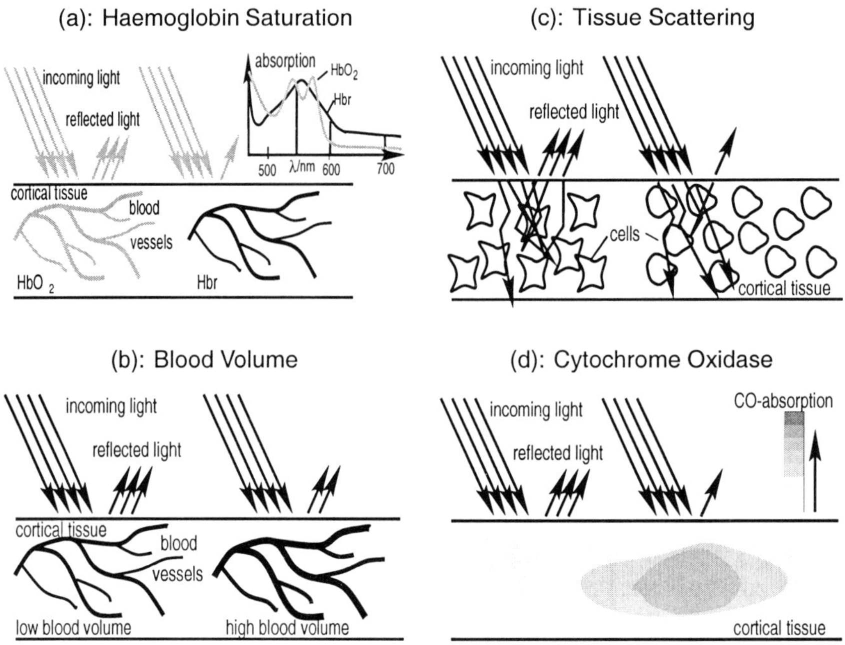


*Figure 4.2.* Biophysical components of intrinsic signals. In each subplot, the left scene corresponds to inactive neurons, the right scene to active neurons. **(a)** Active neurons locally decrease the oxygen saturation of haemoglobin. The tissue reacts by globally increasing the blood-flow and consequently the haemoglobin saturation **(b)** In reaction to neuronal activity, also the total amount of blood in the tissue (blood-volume) increases. **(c)** Active tissue decreases its light scattering coefficient and becomes slightly more transparent. **(d)** Various chromophores (cytochrome oxidase shown) change their spectral behavior depending on the metabolic activity of the tissue.

## 1.3 Optical Imaging of Intrinsic Signals

Optical Imaging of intrinsic signals uses the fact, that local neural activity causes changes of both the light absorption and the light scattering properties of the neural tissue (Bonhoeffer and Grinvald, 1996). These changes are not induced by particular dyes applied to the tissue, but are consequences of the metabolic reaction of the tissue itself to neural activity and are therefore called intrinsic signals. Optical imaging of intrinsic signals offers the important advantage of a relatively weak interaction with the brain tissue under investigation. There is neither phototoxicity nor an interference of dyes with the metabolic state of the cells. However, not all types of intrinsic signals accurately reflect brain activity, and we face the problem of separating the signal components, which are closely bound to neural activity, from other components which we may consider as biological background or noise signals in the context of optical imaging. In chapter 5 – 8 we will provide a detailed treatment of this issue.

Figure 4.2 summarizes some of the major biophysical origins of intrinsic signals (Cohen et al., 1968; Cohen, 1973). When nerve cells in a local cluster fire spikes, the energy consumption of their synaptic terminals increases (Roland, 1993; Magistretti and Pellerin, 1999; Logothetis et al., 2001), and as a consequence the oxygen demand of the tissue that contains these synapses becomes higher. Oxygen diffuses from nearby capillaries to these sites which causes a decrease in the oxygen saturation of the blood (figure 4.2a). Diffusion-based blood deoxygenation in the capillary bed is strongly localized to neural synaptic activity (Logothetis et al., 2001; Almeida and Stetter, 2001). Its spatial extent around the active neuron population is related to the diffusion coefficient for oxygen and to the typical distances of capillaries and ranges around 100 - 200 $\mu$m (Bonhoeffer and Grinvald, 1996). At the same time, the amount of light scattering close to the active neurons decreases (figure 4.2c). Although the detailed mechanisms that relate scattering to neural activity are still unclear, it could be shown that the optical transmittance of hippocampal tissue in vitro increases both after induced cell swelling (Kreisman et al., 1995) and neural excitation (MacVicar and Hochman, 1991; Andrew and MacVicar, 1994), and that these reactions are localized to regions spred approximately 100 - 300 $\mu$m around the active tissue (Orbach et al., 1985; MacVicar and Hochman, 1991).

Once the oxygen level in the tissue decreases, the tissue reacts by providing additional oxygen using two mechanisms: An increase in blood-flow globally increases the oxygen saturation of blood far above its resting level, and simultaneously the amount of blood in the tissue, the blood volume, increases (figure 4.2b). Changes in blood-flow and blood-volume extend by more than 500 $\mu$m (Bonhoeffer and Grinvald, 1996). Finally, several chromophores including cytochrome oxidase (figure 4.2d) and NADPH change their absorption characteristics dependent on the oxygenation state of the tissue. All intrinsic signals are slow compared to the temporal change in neuronal activity. Usually, their onset is delayed by $0.3 - 3$ sec with respect to the stimulus onset and they usually reach their maximum after several seconds.

The biophysical mechanisms summarized above cause changes in the optical properties of the tissue, which we refer to as intrinsic signals. Intrinsic signals modify the light reflectance of tissue, which can be detected by a CCD-camera (figure 4.1). Figure 4.3 summarizes the relationship between different intrinsic signals (middle column), their biophysical origins (left column), and the different components we find in the images of the brain surface (right column). We neglect the influence of chromophores, the contribution of which to intrinsic signals is still unclear.

The most important intrinsic signals are localized changes in light absorption and scattering, because they reflect local neuronal activity. The local change in absorption is probably mostly carried by diffusion-based blood-deoxygenation though direct evidence for this relationship is still lacking (Mayhew et al., 1999).

## 2. Monte-Carlo Simulation of Optical Imaging

Cortical tissue represents an optically turbid medium that for red and near infrared wavelengths is characterized by strong light scattering (scattering coefficients $\mu_s \approx 10\text{mm}^{-1}$) with a strong forward scattering anisotropy ($g \approx 0.9$), and a weak absorption with absorption coefficients near $\mu_a \approx 0.1\text{mm}^{-1}$) (Cheong et al., 1990). As a consequence, photons entering the tissue typically undergo multiple mostly forward directed scattering processes before they are either absorbed or remitted from the tissue. Therefore, photons can propagate to considerable depths and a significant fraction of them can spread over considerable lateral distances of more than 1 mm, before they are remitted. Intrinsic signals are sampled through the whole volume passed by the photons. Consequently, information about the depth of these signals can not be detected using a CCD camera system. Further, due to the lateral spread of the photon pathways, intrinsic signals are strongly blurred by the tissue, which leads to a loss both in lateral resolution and contrast. Under these conditions, it is not necessarily intuitive, if and under which conditions small intrinsic signals can be reliably detected with sub-millimeter accuracy.

In order to answer this question, we setup a model cortical tissue that is described by the optical properties measured for human gray matter in vitro (Cheong et al., 1990). Because the role of scattering changes for intrinsic signal imaging is still not completely clear, we concentrate on absorption signals. In the model, a local strongly absorbing sphere is embedded into the tissue and serves as a model for an intrinsic absorption signal or a dye absorption signal caused by a localized cluster of active neurons. Into this model tissue we inject a narrow beam of photons and calculate their pathways through the tissue using Monte Carlo simulations (Graaf et al., 1993a; Graaf et al., 1993b; Schmitt and Ben-Letaief, 1996). From all photons that exit the model tissue we obtain the diffuse reflectance pattern for this illumination condition. These simulations are carried out for different locations of the incoming beam, and from the superposition of all reflectance patterns we obtain a simulated image of the absorption signal.

### 2.1 Light Propagation in Inhomogeneous Turbid Media

As a model of the grey matter of the neocortex, we consider a semi-infinite continuous turbid medium which extends to infinity in the $x$- and $y$-directions and in the negative $z$-direction. Its surface lies within the $x$-$y$-plane (see also figure 4.6). The optical properties of this medium can be characterized by its index of refraction $n$ (assumed to be constant), its absorption coefficient $\mu_a(\mathbf{r})$, its scattering coefficient $\mu_s(\mathbf{r})$, and by a phase function $p(\mathbf{r}, \mathbf{s}, \mathbf{s}')$, where $\mathbf{r} = (x, y, z)$ denotes the position in space and $\mathbf{s}, \mathbf{s}'$ represent unit vectors describing the propagation direction of a photon before and after a scattering

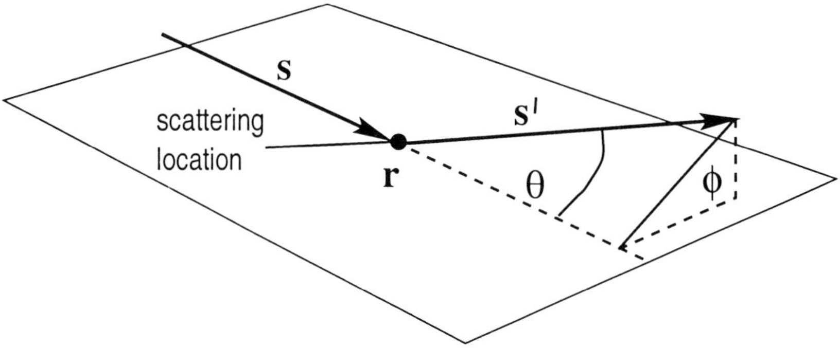


*Figure 4.5.* The phase function $p(\mathbf{r}, \mathrm{s}, \mathrm{s}')$ describes the probability density for a photon to be scattered from propagation direction s into direction $\mathrm{s}'$ at $\mathbf{r}$. For isotropic media, the phase function depends only on the position $\mathbf{r}$ and on the scattering angle $\theta$.

process (Cheong et al., 1990; Klier, 1972; van Gemert et al., 1989; Wilson and Jacques, 1990). The coefficients $\mu_a(\mathbf{r})$ and $\mu_s(\mathbf{r})$ are the probability density functions for an absorption or a scattering event of a photon at the location $\mathbf{r}$, while the phase function $p(\mathbf{r}, \mathrm{s}, \mathrm{s}')$ is the probability density function for scattering a photon at the location $\mathbf{r}$ from direction s into direction $\mathrm{s}'$ (figure 4.5).

In order to describe the propagation of photons through the tissue, expressions for scattering and absorption probabilities along given trajectories are required. In order to obtain these quantities, we consider a bundle of $n_0$ photons, which propagate from a starting location $\mathbf{r}_0$ along a straight trajectory $\mathbf{r}(t) = \mathbf{r}_0 + l\ \mathrm{s},\quad l \geq 0$. In this formulation, $l$ is identical to the distance of the photons from the starting point. Due to absorption and scattering processes, the number of photons along the trajectory decreases according to

$$\frac{dn(l)}{dl} = -\mu(\mathbf{r}(l))\ n(l), \tag{4.1}$$

while the photons propagate through a layer of tissue between $\mathbf{r}_0 + l\mathrm{s}$ and $\mathbf{r}_0 + (l + dl)\mathrm{s}$. The coefficient $\mu$ represents $\mu_a$ for absorption and $\mu_s$ for scattering respectively. If absorption and scattering processes are statistically independent, the total attenuation due to both mechanisms results as

$$\frac{dn(l)}{dl} = -\left(\mu_a(\mathbf{r}(l)) + \mu_s(\mathbf{r}(l))\right)\ n(l) =: -\mu_t(\mathbf{r}(l))\ n(l) \tag{4.2}$$

where $\mu_t(\mathbf{r}) = \mu_a(\mathbf{r}) + \mu_s(\mathbf{r})$ is the total attenuation coefficient. Integration of the differential equations (4.1) or (4.2) along the trajectory yields

$$n(l) = n_0 \exp\left(-\int_0^l \mu(\mathbf{r}(l'))\ dl'\right) \tag{4.3}$$

with $\mu = \mu_a, \mu_s$ or $\mu_t$ respectively. If the ensemble of initially $n_0$ photons propagates about a distance of $l_0$, $n(l_0)$ of them reach the point $\mathbf{r}(l_0)$ without interacting with the tissue. Thus, in the limit of large $n_0$, one obtains the 'total survival probability' for a photon along the trajectory as

$$P_t(l) = \lim_{n_0 \to \infty} \frac{n(l)}{n_0} = \exp\left( - \int_0^l \mu_t(\mathbf{r}(l'))\, dl' \right). \tag{4.4}$$

$P_t(l)$ represents the probability for a photon not to undergo any tissue interaction between $\mathbf{r}_0$ and $\mathbf{r}_0 + l\mathbf{s}$. Similarly, the survival probability under consideration of absorption events only becomes

$$P_a(l) = \exp\left( - \int_0^l \mu_a(\mathbf{r}(l'))\, dl' \right). \tag{4.5}$$

and the survival probability for scattering is

$$P_s(l) = \exp\left( - \int_0^l \mu_s(\mathbf{r}(l'))\, dl' \right). \tag{4.6}$$

Note, that due to the inhomogeneity of the medium, the probabilities (4.4) - (4.6) depend on the trajectory (i.e. on $\mathbf{r}_0$ and $\mathbf{s}$) of the photons within the tissue. In the special case of a homogeneous medium, $\mu_a(\mathbf{r}) = \mu_a, \mu_s(\mathbf{r}) = \mu_s$, the interaction probabilities reduce to simple exponential expressions,

$$P_a(l) = \exp(-\mu_a\, l), \tag{4.7}$$

$$P_s(l) = \exp(-\mu_s\, l), \tag{4.8}$$

which in case of absorption are well-known as Beer's law.

If a photon is scattered after its free path length, the new propagation direction is determined according to the phase function $p(\mathbf{r}, \mathbf{s}, \mathbf{s}')$. If the medium is isotropic, the phase function at a fixed scattering location depends only on the scattering angle $\theta$, $p(\mathbf{r}, \mathbf{s}, \mathbf{s}') \equiv p(\mathbf{r}, \theta)$, where $\theta$ is defined as the angle between the propagation direction before and after the scattering process, $\cos(\theta) = \mathbf{s}^T \cdot \mathbf{s}'$, and "$\cdot$" denotes the inner product of two vectors. The azimuth angle $\phi$ around the axis defined by $\mathbf{s}$ (figure 4.5) is then uniformly distributed. For many applications, the detailed dependence of $p(\mathbf{r}, \theta)$ on the scattering angle $\theta$ is not critical (Wilson and Jacques, 1990) (nor is it known), and it is adequate to characterize it by its first moment, the anisotropy parameter $g(\mathbf{r})$ with

$$g(\mathbf{r}) = \int_0^\pi p(\mathbf{r}, \theta) \cos(\theta) d\theta. \tag{4.9}$$

In Monte Carlo simulations, where a phase function for the scattering angle is needed, one convenient choice for the phase function is the Henyey-Greenstein-function (Schmitt and Ben-Letaief, 1996; Henyey and Greenstein, 1941; Flock et al., 1989a),

$$p(\mathbf{r},\theta) = \frac{1}{4\pi}\frac{1-g(\mathbf{r})^2}{\left(1+g(\mathbf{r})^2-2g(\mathbf{r})\cos(\theta)\right)^{3/2}}. \tag{4.10}$$

In the special case of a homogeneous medium, the light propagation in biological tissue (in the limit $\mu_s \gg \mu_a$) can be described by diffusion theory yielding an effective penetration depth of (Wilson and Jacques, 1990)

$$\delta_{eff} = \left(3\mu_a(\mu_a + \mu_s(1-g))\right)^{-1/2}. \tag{4.11}$$

The diffuse reflectance pattern $D(\sigma)$ caused by illumination with a narrow, vertically incident light beam is then

$$D(\sigma) = \frac{1}{\sigma^2}\exp(-\sigma/\delta_{eff}), \tag{4.12}$$

where $\sigma$ denotes the two-dimensional distance vector between a location on the surface of the tissue and the incoming light beam. For example, biologically plausible parameters at 633 nm light (i.e. those of human gray matter in vitro: $\mu_a = 0.26\ \text{mm}^{-1}$, $\mu_s = 6.02\ \text{mm}^{-1}$, and $g = 0.88$ (Cheong et al., 1990)) lead to an effective penetration depth of $\delta_{eff} = 0.875$ mm: Incident photons propagate approximately 0.9 mm into the tissue and laterally spread by the same amount. In section 3, we will provide results about the detailed behavior of the penetration depth.

## 2.2 Simplified Tissue Model for Localized Absorbers

Localized intrinsic blood-related signals are caused by a local increase of the absorption coefficient of cortical tissue at red and near-infrared light due to local deoxygenation of blood by active neurons. Our goal is to determine the pattern of diffusely remitted light (i.e. the intrinsic blood-related signal) under various illumination conditions in the presence of a local absorber within the tissue compared to the case without that absorber. We model this situation by considering the simple case of an isotropic semi-infinite medium, i.e. anisotropies such as the mainly vertical apical dendrites of pyramidal neurons or horizontal nerve fiber bundles from pyramidal collaterals (Rockland and Lund, 1983) are neglected in the present simple approach. This model tissue is taken to have constant optical parameters $\mu_a$, $\mu_s$, and $g$ everywhere except within a hard sphere with radius $\rho$, which is centered at a depth $-Z < 0$ below the origin of the coordinate system, i.e. the center of which is located at $\mathbf{R} = (0, 0, -Z)$ (figure 4.6). This sphere serves as a model for the local variation of the light absorption within the tissue (i.e. a blood-related intrinsic signal), which could be evoked by a local neural activity center. According to that, the absorption coefficient within the sphere is increased by $\mu_A > 0$ resulting in a total absorption

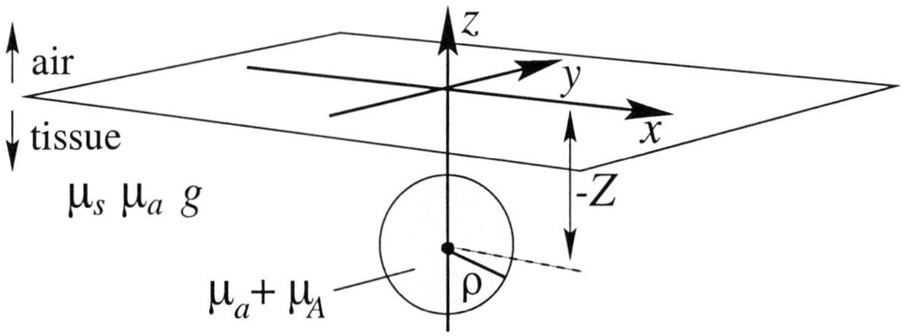


*Figure 4.6.* The model tissue is assumed to be an isotropic, semi-infinite medium with constant optical parameters $\mu_a$, $\mu_s$, and $g$. Within a small sphere with radius $\rho$ centered around $\mathbf{R} = (0, 0, -Z)$ the absorption coefficient is increased by $\mu_A$. This spheric absorber models a localized blood-related intrinsic signal source caused by a small neural activity blob.

coefficient of

$$\mu_{a,\text{sphere}} = \mu_a + \mu_A \tag{4.13}$$

for the local absorber. $\mu_A$ will be referred to as incremental absorption coefficient. The scattering properties, in contrast, are assumed to be identical within and outside the sphere.

Within this framework, the scattering probability for a photon propagating within the tissue simply reduces to Beer's law, Eq. (4.8) and the phase function is given by a single Henyey-Greenstein expression, $p(\mathbf{r}, \theta) \equiv p(\theta)$ eq. (4.10). The only quantity that depends on the trajectory of a photon through the tissue is the absorption probability $P_a(l)$ eq. (4.5). It can be integrated analytically for a given photon trajectory $\mathbf{r}(l) = \mathbf{r}_0 + l\mathbf{s}, \quad l \geq 0$. For that trajectory we determine the intersection points $l_1$ and $l_2$ of the trajectory with the boundary of the absorbing sphere. They are given as the solution of the equation

$$|l\mathbf{s} - (\mathbf{R} - \mathbf{r}_0)|^2 = \rho^2, \tag{4.14}$$

which can be written as

$$l_{1,2} = \mathbf{s}^T \cdot \mathbf{\Delta r} \pm \sqrt{(\mathbf{s} \cdot \mathbf{\Delta r})^2 - (\Delta r^2 - \rho^2)}, \tag{4.15}$$

where $\mathbf{\Delta r} = \mathbf{R} - \mathbf{r}_0$, $\Delta r = ||\mathbf{\Delta r}||$ and a real solution is only defined for a positive argument of the square root. The solutions of Eq. (4.15) belong to three different spatial scenarios (figure 4.7): $(i)$ The whole trajectory runs outside the sphere or touches it in only one point (two negative solutions, one or no real solution), $(ii)$ the trajectory starts inside the sphere and leaves it (one positive and one negative solution), or $(iii)$ it starts outside the sphere, enters it and leaves it again (two positive solutions). Now the integration within the

exponent of the survival probability for absorption, Eq. (4.5), can be carried out for the three cases leading to the following probabilities:
Case $(i)$, trajectory outside the sphere:

$$P_a(l) = \exp(-\mu_a\, l). \tag{4.16}$$

Case $(ii)$, trajectory starts inside the sphere:

$$P_a(l) = \begin{cases} \exp(-\mu_a\, l)\ \exp(-\mu_A\, l) & l < l_1 \\ \exp(-\mu_a\, l)\ \exp(-\mu_A\, l_1) & l \geq l_1 \end{cases} \tag{4.17}$$

Case $(iii)$, trajectory penetrates sphere:

$$P_a(l) = \begin{cases} \exp(-\mu_a\, l) & l < l_1 \\ \exp(-\mu_a\, l)\ \exp(-\mu_A\, (l - l_1)) & l_1 \leq l < l_2 \\ \exp(-\mu_a\, l)\ \exp(-\mu_A\, (l_2 - l_1)) & l \geq l_2 \end{cases} \tag{4.18}$$

Hence, the total probability $P_a(l_0)$ for a photon to survive the propagation from $\mathbf{r}_0$ to $\mathbf{r}_0 + l_0\mathbf{s}$ without being absorbed is given by the product of elementary survival probabilities, where each single probability describes only the contribution of the propagation through the homogeneous medium ($\exp(-\mu_a\, l_0)$) or through the sphere (for example $\exp(-\mu_A\, (l_2 - l_1))$). This property, which simply arises from the assumed independence of different absorption mechanisms, allows for a straightforward generalization of the calculation of $P_a(l_0)$ to several absorbing regions $S_1, ..., S_M$ with incremental absorption coefficients $\mu_{A,1}, ..., \mu_{A,M}$, which can be efficiently numerically implemented: One determines the path length $\Delta l_m$ of the photon within each absorber $S_m$, which then yields the survival probability

$$P_a(l_0) = \exp(-\mu_a\, l_0) \prod_{m=1}^{M} \exp(-\mu_{A,m}\, \Delta l_m). \tag{4.19}$$

## 2.3 Monte-Carlo Simulations

Monte-Carlo methods for the description of light propagation through turbid media have been described in (Graaf et al., 1993a; Graaf et al., 1993b; Schmitt and Ben-Letaief, 1996; Flock et al., 1989a; Flock et al., 1989b). We apply the method in order to characterize the propagation pathways of photons through the model tissue specified above. In order to achieve that, a number of photons is vertically injected into the medium, and their multiply scattered pathways through the medium are followed (up to a maximum number of iterations) until they are either absorbed or leave the medium again.

The propagation pathway for each photon through the tissue is determined as follows. First the photon is injected into the medium at a fixed position $(x_I, y_I)$

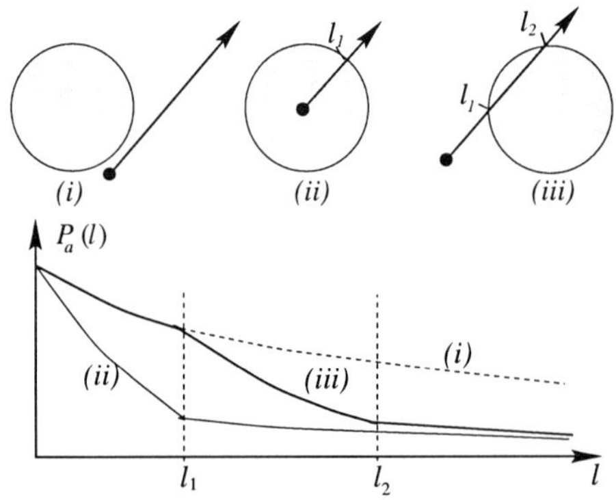


*Figure 4.7* Behavior of the survival probability $P_a(l)$ of a photon along a trajectory $\mathbf{r}_0 + l\mathbf{s}$ for the three practically different cases, namely: $(i)$ the photon misses the sphere, $(ii)$ it leaves the sphere, and $(iii)$ it penetrates the sphere.

and initially (at time step $t = 0$, i.e. before its first tissue interaction) follows the trajectory $\mathbf{r}_0(t = 0) + l\mathbf{s}(t = 0)$ with $\mathbf{r}_0(0) = (x_I, y_I, 0)$ and $\mathbf{s}(0) = (0, 0, -1)$. In order to determine the location of the next tissue interaction, the propagation length $l_s$ is drawn from the survival probability $P_s(l)$, eq. (4.8), which provides the current free path length in presence of tissue scattering alone. Next, the probability for the photon to be absorbed along its current free pathway, i.e. between $\mathbf{r}_0$ and $\mathbf{r}_0 + l_s\mathbf{s}$, is determined. This probability is just $1 - P_a(l_s)$, where $P_a(l)$ is taken from eqs. (4.16)-(4.18) using $\mathbf{r}_0(t)$ and $\mathbf{s}(t)$. Comparison of a uniformly distributed random number $x$ (between 0 and 1) to $P_a(l_s)$ decides, whether the photon is absorbed ($x > P_a(l_s)$) or not. If the photon survives, the start of its next trajectory is taken to be $\mathbf{r}_0(t+1) = \mathbf{r}_0(t) + l_s\mathbf{s}(t)$, and its next propagation direction $\mathbf{s}(t+1)$ is determined from the scattering angle $\theta$ and $\phi$ (see figure 4.5), which in turn are drawn from the Henyey-Greenstein distribution eq. (4.10) and from a uniform distribution between 0 and $2\pi$ respectively.

For the detection of exiting photons, the tissue surface was divided into a raster of $(2n_a + 1) \times (2n_a + 1)$ squares with diameters $\Delta a$ around the photon injection site, which was kept constant for each single simulation. The number of exiting photons and their properties were stored separately for each square and were assigned to the corresponding raster point. For each square $(x, y)$, the photons were sorted according to $n_z$ different maximum penetration depths in steps of $\Delta z$ and $n_\alpha$ different exit angles in steps of $\Delta\alpha = \pi/(2n_\alpha)$ (figure 4.8).

If a photon exits the medium, it is first tested, if it undergoes total reflection on the surface of the medium. If this is the case, the photon trajectory is reflected back into the medium, otherwise the photon is allowed to exit the medium where the exit direction is determined according to Snell's law. Reflection processes besides total reflection are not taken into account, because during the analysis of the remitted patterns we consider only photons exiting steeper than 60 deg,

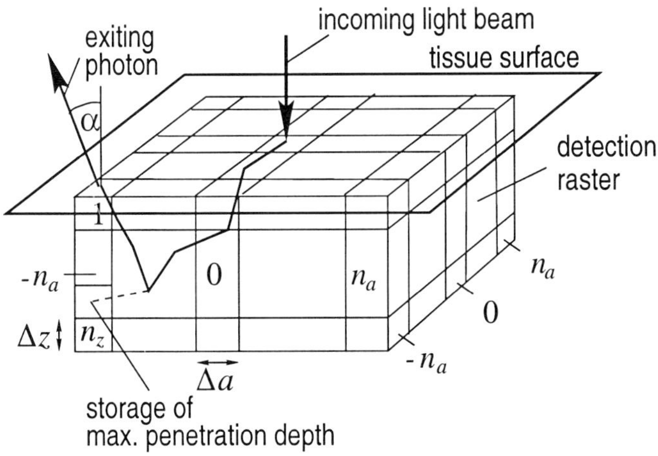


*Figure 4.8.* Detection of the diffuse reflectance pattern. The region around the photon injection point is divided into a raster of $(2n_a + 1) \times (2n_a + 1)$ pixels of $\Delta a$ on a side. If a photon exits at a given square, its maximum penetration depth is stored in a corresponding penetration depth histogram with $n_z$ bins and penetration depth intervals $\Delta z$. Also, the exit angle $\alpha$ between the surface normal and its exit direction is stored.

where the specular reflection is weak and not very strongly dependent on the exit angle. This maximum exit angle for photons to be detected corresponds to a numerical aperture of $NA = 0.5$ of the considered imaging system. For each exiting photon, the maximum penetration depth, the exit location and the exit angle between the trajectory and the surface normal are stored. After the injection of a large amount of photons, these quantities characterize the diffuse reflectance patterns under the considered illumination condition.

The simulator was implemented in C and TCL/TK on SUN Sparc 4 and Sparc 20 workstations. All simulations were run using a single set of optical parameters, that were taken to be those of human gray matter in vitro at 633 nm, i.e. $\mu_a = 0.26$ mm$^{-1}$, $\mu_s = 6.02$ mm$^{-1}$, and $g = 0.88$ (Cheong et al., 1990). The index of refraction was taken to be that of water, i.e. $n = 1.33$. If a local absorbing sphere was present, its radius was $\rho = 0.1$ mm and its incremental absorption coefficient was set to $\mu_A = 10$ mm$^{-1}$. This possibly unrealistically high value was chosen, because it does not change any of our results qualitatively (see also section 4.5) but strongly increases the quantitative effect of the local absorber on the simulated diffuse reflectance pattern.

## 3. Results for Simulated Video Imaging

Figure 4.9a displays a tangential view on the simulated trajectories of 600 photons within the tissue (below x-axis). While the effective penetration depth

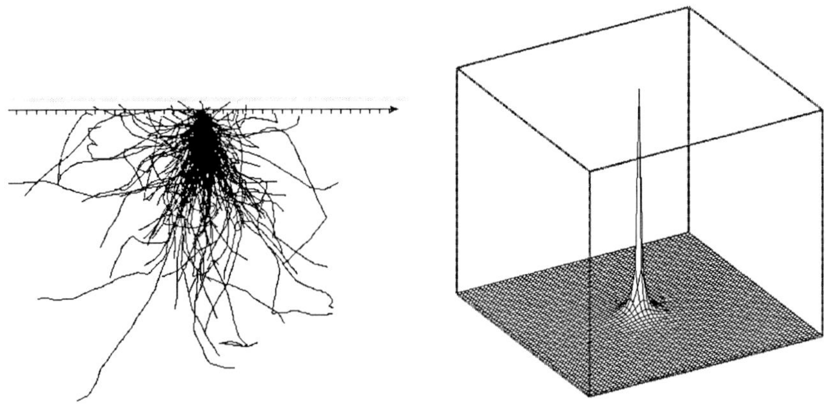

(a) (b)

*Figure 4.9.* **(a)** Simulated trajectories for 600 photons injected within a model tissue (view tangential to tissue surface, which coincides with the x-axis). Units: 500 $\mu$m per tick. Photons can enter several centimeters into the tissue. **(b)** Diffuse reflectance pattern calculated for $N = 10^8$ photons injected into the tissue. Grid constant of the mesh: 50 $\mu$m. For the parameters used, the reflectance pattern extends to approximately 1 mm ($\delta_{eff} = 0.875$ mm). Optical parameters: $\mu_a = 0.26\text{mm}^{-1}$, $\mu_s = 6.02\text{mm}^{-1}$, and $g = 0.88$)

amounts to less than 1 mm, some photons propagate many centimeters through the tissue before they become absorbed or remitted. Figure 4.9b shows a calculated reflectance pattern for $N = 10^8$ photons injected at a single spot at the origin of a homogeneous model tissue. For the parameters used, the reflectance pattern spreads approximately 1 mm to each side. Both plots demonstrate, that $(i)$ locally injected photons sample a large volume of tissue in the range of several cm$^3$, before they exit the tissue again, and $(ii)$ a considerable part of local the image of signals can theoretically be spread up to a millimeter caused by the lateral extent of the diffuse reflectance pattern. Under these conditions, it is not necessarily intuitive, if and under which conditions small intrinsic signals with sub-millimeter sizes can be reliably imaged by a CCD-camera. In the following we will therefore treat this issue quantitatively.

## 3.1 Lateral Resolution and Contrast for Video Imaging

In conventional optical imaging using a CCD-camera (henceforth referred to as video imaging), the brain surface is illuminated by a collimated light beam. We use an idealized description of this beam and think of it as a grid of parallel and infinitely sharp photon rays, which hit the model tissue vertically on the vertices of a square lattice. Each of these incident photon rays generates a diffuse reflectance pattern, and the CCD-camera forms an image of the brain surface by simultaneously recording the superposition of all these patterns. Consequently, we simulate standard optical imaging by defining a two-dimensional photon injection raster $(x_{I,m}, y_{I,n}) = (m\Delta r, n\Delta r), \quad m, n = 0, \pm 1, ..., \pm M$ on the surface of the model tissue. At each of the $(2M + 1)(2M + 1)$ raster points, $N$ photons are vertically injected into the tissue, the diffuse reflectance patterns are calculated and stored as described in the previous section. For conventional video-imaging, the tissue surface is covered with a single large raster. For each pixel $(x, y)$ of this raster, the photons of all $(2M+1)(2M+1)$ reflection patterns emerging through that pixel are summed up yielding the distribution $I(x, y)$ of remitted photons after the application of all $(2M + 1)(2M + 1)$ incoming light beams. This corresponds to a simultaneous illumination of all raster points with in total $(2M + 1)(2M + 1)N$ photons and a simultaneous detection of the remitted light. In order to avoid boundary effects, i.e. in order to take into account photons that are injected outside the injection raster but exit the tissue within the detection raster, injection points outside the injection raster are added and equipped with the reflectance pattern obtained for the homogeneous medium figure 4.9.

Now we calculate the simulated CCD-image of a model tissue, that is equipped with a local absorber (radius $\rho = 0.1$mm, incremental absorption coefficient $\mu_A = 10\text{mm}^{-1}$) centered below the origin at $\mathbf{R} = (0, 0, -Z)$ as shown in figure 4.6). In order to determine the lateral resolution and the contrast, we carried out the following simulations: $N = 10^6$ photons were injected into each of $21 \times 21$ illumination points located on a square grid (grid constant 0.05mm) around the origin covering a total area of $1 \times 1$mm. No penetration depth information was resolved, i.e $n_z = 1, \Delta z = \infty$. The grid constant of the detection raster was taken to be the same than that of the photon injection raster. The resulting image $I(x, y)$ for conventional optical imaging using a video- or CCD-camera was computed as described above.

Figure 4.10 shows the simulated images of the absorber if it is located at a depth of $Z = 0.2$ and $0.6$ mm respectively. Even for a very shallow location of the absorber, its video image is already strongly blurred (left), while for $Z = 0.6$ mm it is not detectable anymore in the current framework of $10^6$ injected photons. The figure demonstrates that though photons can propagate very far into the tissue (figure 4.9a), the changes in reflectance detected on the surface of the tissue originate almost exclusively from absorption signals within the superficial 500 $\mu$m of the tissue.

In order to quantify the relative contribution of absorption signals in various depths to the total reflectance of the tissue, we calculated CCD-images for different absorber depths between 0.1 and 1 mm and calculated the contrast $c$ of the resulting images $I(x, y)$ as

$$c = \frac{\max_{x,y}(I(x, y)) - \min_{x,y}(I(x, y))}{\max_{x,y}(I(x, y))}. \tag{4.20}$$

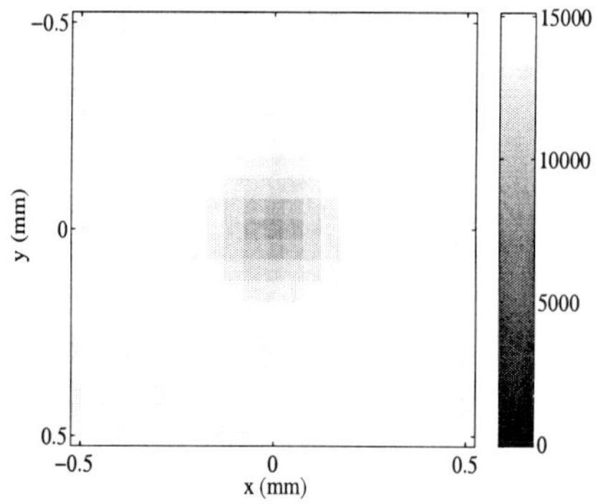


absorber depth: $Z = 0.2$ mm

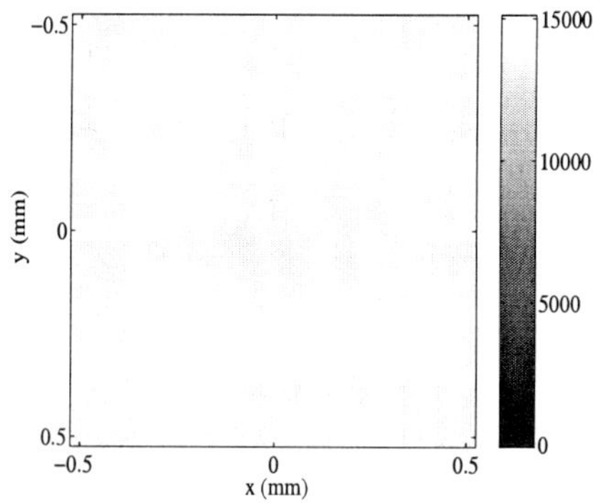


absorber depth: $Z = 0.6$ mm

*Figure 4.10.* Simulated video images for a homogeneous tissue with a local absorber at $Z = 0.2$ mm (left) and $Z = 0.6$ mm (right). The relative strength of the signal strongly decreases with depth, i.e. most of the change in reflectance originates from absorbers closer than 600 $\mu$m to the surface. Gray scale: Number of remitted photons.

Figure 4.11a shows one-dimensional slices of the simulated images of the absorbing sphere located at depths of 0.1 mm (solid), 0.3 mm (dashed) and 0.5 mm (dash-dotted) below the surface. The vertical dotted lines mark the actual radius of the absorbing sphere. The figure demonstrates, that the images of the absorber become blurred by the influence of strong tissue scattering and extend beyond the true radius of the imaged sphere (dotted lines). The blur, which is depth-dependent and amounts between approximately 50 $\mu$m for $Z = 0.1$ mm and 100 $\mu$m for a depth of $Z = 0.5$ mm to either side. Hence, the image of a point-like intrinsic signal at a depth between 0.2-0.5 mm spreads to approximately 200 $\mu$m. This result agrees well with experimental results from hippocampal slices (Kreisman et al., 1995).

The solid line in figure 4.11b quantifies the contrast of the images according to eq. (4.20) as a function of the depth of the absorber. The image contrast strongly decreases and between 0.3 and 0.5 mm saturates at a low level, which is related to noise variability. Some aspects of cortical function are organized in a columnar fashion. Therefore, one may ask, which parts of a vertical and columnar intrinsic absorption signal contributes to the video image. This question can be addressed in the present framework, if we approximate a columnar absorber as a sequence of spherical absorbers located at the same $(x, y)$ position but at increasing depths. If the column reaches from $Z = 0$ up to a depth of $Z_0$, its image is approximately given by the sum of the images of the individual spheres located between the two depths: $c_{\mathrm{cum}}(Z_0) = \sum_{Z=0.1}^{Z_0} c(Z)$. Note that this is only true, if the superposition principle holds, which is likely to be given for realistic changes in absorption. The cumulative contrast $c_{\mathrm{cum}}(Z_0)$, normalized to its

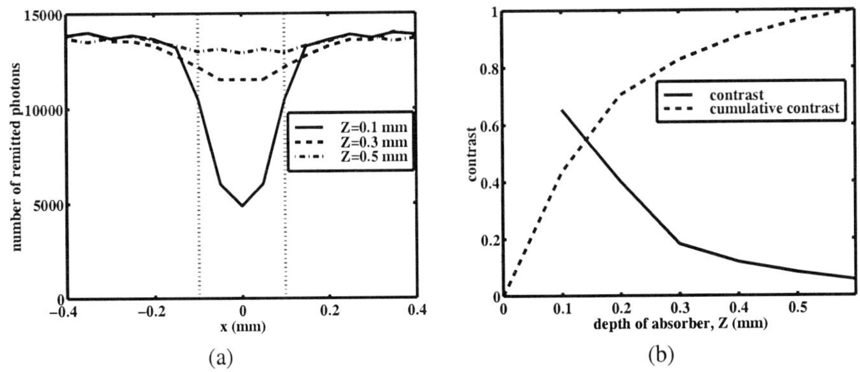


*Figure 4.11.* Image contrast for video imaging. **(a)** 1D-slices through the images of a spherical absorber 0.1 mm (solid), 0.3 mm (dashed) and 0.5 mm (dash-dotted) below the surface. Vertical dotted lines mark the true absorber radius of 0.1 mm. **(b)** Solid line: Image contrast $c$ from eq. (4.20) as a function of absorber depth. Dashed line: Cumulative signal strength of a columnar structure obtained as the normalized integral over the contrast. Images of intrinsic signals originate from the superficial 0.3 - 0.5 mm of the nerve tissue.

value at $Z = 0.6$ mm, is shown as dashed line in figure 4.11b. The curve strongly increases up to 0.2 mm and then starts to saturate. This demonstrates that the image of a columnar intrinsic signal is formed mainly by its top 0.2 - 0.5 mm.

## 3.2 Detectability Threshold

In the simulations above, the number of injected photons per illuminated spot was arbitrarily chosen as $N = 10^6$ photons. Since optical imaging works under strong illumination conditions, the noise sources for the detection of changes in reflectance are dominated by photon shot noise ranging around $\sqrt{N}$, which in turn restricts the signal-to-noise ratio for the detection of these reflectance changes. In this estimate, we treat biological noise, i.e. reflectance changes that are not related to neural activity (such as slow changes in the oxygenation level of blood (Bonhoeffer and Grinvald, 1996)), as signal rather than as noise, as they are generated by the tissue and not by the physical recording conditions. Photon shot noise levels can be estimated by calculating the approximate number of photons that are used to record a single pixel during a scan. For a wavelength of 605 nm (which is frequently taken for optical imaging (Bonhoeffer and Grinvald, 1996)), a single photon carries roughly the energy of $E_p = hc/\lambda = 3 \times 10^{-19}$J ($h$ is Planck's constant and $c$ is the speed of light). A rough estimate for the photo damage limit for biological tissue (though estimated for skin only) is 0.3W/cm$^2$ for illuminations up to 100s (Sliney and Wolbarsht, 1980). If the scanning light beam covers a surface of 5mm $\times$ 5mm, a realistic irradiation power for that beam is 10mW, which leads to an intensity roughly ten times below the photo damage limit. For that power, the light beam carries about $3 \times 10^{16}$ photons per second. If the whole image is divided into $256 \times 256$ pixels, each pixel is then recorded using roughly $N_t = 4.6 \times 10^{11}$ photons at a 1 Hz frame rate. Hence, the noise level in this experiment would be reduced by a factor $\sqrt{N/N_t}$ compared to the shown simulation results, i.e. approximately 1/680 for the simulations of the previous section. The background level of reflected light in the images is about 14000 photons per pixel in the simulations (figure 4.11a), which in the experiment described above would be increased to $N_p = 6.4 \times 10^9$ backscattered photons. The minimum detectable signal is then $\sqrt{N_p} = 80000$ photons and the minimum detectable contrast becomes $\sqrt{N_p}/N_p = 1.25 \times 10^{-5}$. For a very coarse estimation of the minimum detectable change in the absorption coefficient, we assume a linear relationship between the strength of the incremental absorption coefficient $\mu_A$ within the sphere and the resulting contrast within the image (i.e. linearization of the exponential absorption probability). In this case, the simulated contrast caused by an absorber with $\mu_A = 10\text{mm}^{-1}$ located at a depth of 0.3mm is $c = 0.18$ (figure 4.11b). The estimated minimum detectable change in absorption coefficient for this depth becomes $\Delta\mu_A^{min} \approx (10 \times 1.25 \times 10^{-5}/0.18)\ \text{mm}^{-1} \approx 7\ 10^{-4}$ $\text{mm}^{-1}$.

This value has to be compared with a realistic change in the absorption coefficient $\mu_A$ as caused by a blood-related mapping signal. This value can be estimated as follows: Human blood contains approximately $c_t = 150$ g/l of haemoglobin (Schmidt and Thews, 1989). With a molecular weight of 65400 for haemoglobin, this value translates to a molar concentration of $cb_{Hb} = 2.29$ mmol/l. Because about 5 percent of brain tissue consists of blood, the concentration of haemoglobin in brain tissue becomes $c = 0.05cb_{Hb} \approx 0.11$ mmol/l. At 600 nm light wavelength, the extinction coefficient for deoxygenated haemoglobin (Hbr) is $\varepsilon_r = 0.35$ l/(mmol mm), and the coefficient for oxygenated haemoglobin ($HbO_2$) is $\varepsilon_o = 0.1$ l/(mmol mm). At rest, a fraction $x$ of the haemoglobin is oxygenated, and the rest is deoxygenated. $x$ is called the saturation level. The absorption coefficient due to the blood in brain tissue becomes

$$\mu_a^b = \ln(10)c(x\varepsilon_o + (1-x)\varepsilon_r) \tag{4.21}$$

At rest, the blood saturation level in parenchyma is approximately $x = 0.5$, and is changed by 2 percent by neuronal activity (Mayhew et al., 2000). According to eq. (4.21), this corresponds to an activity-related change in absorption coefficient of $\mu_A \approx 1.2 \cdot 10^{-3}\ \text{mm}^{-1}$. A comparison with the estimated detection threshold of $\mu_A^{min} \approx 7 \cdot 10^{-4}\ \text{mm}^{-1}$ demonstrates, that video-images of blood-related intrinsic signals are close to the detectability threshold for a single-trial experiment and definitely originate from the superficial 0.4 mm of brain tissue.

# 4. Simulated Scanning Laser Optical Imaging

As we have seen, video imaging records intrinsic signals only from the superficial 400 $\mu$m of the tissue and in addition integrates intrinsic signals over this depth. The cortex, in contrast, is in total roughly 2 mm thick and carries out different processing tasks within different layers. For example, the superficial layers of the macaque monkey striate cortex are dominated by orientation selectivity and color sensitivity of the cells (Hubel and Wiesel, 1968; Hubel and Wiesel, 1977), while direction selective neurons concentrate in layer 4b (Hawken et al., 1988). Also, many of the layer 4C neurons are not or weakly orientation selective and can be distinguished by their color sensitivity and contrast threshold (Hubel and Wiesel, 1977). While the structure of orientation maps within the superficial layers of macaque primary visual cortex has been examined in several studies, the three-dimensional spatial distribution of direction selectivity, color sensitivity and contrast thresholds could not be addressed with the standard technique because these properties may reside in different depths (cf chapter 3, section 2.4). Hence, it would be desirable to find an extension of conventional optical imaging in a way that $(i)$ signal detection can be performed for deeper cortical structures and $(ii)$ signals can be detected selectively for some depth within the cortex.

Confocal scanning laser techniques have been used in ophthalmoscopy (with fast image sampling rates up to video frequency of 50 or 60 frames per second (Webb and Hughes, 1981; Fitzke et al., 1991; Fitzke and Masters, 1993; Stetter et al., 1995d; Stetter et al., 1995e; Timberlake and Stetter, 1996; Stetter et al., 1996; Babel et al., 1997)), in densitometry (Tornow et al., 1997; Marcos et al., 1997) and in confocal microscopy (Pawley, 1995; Scholz et al., 1998; Bucher et al., 2000), where they have been shown to allow for depth-resolved detection of fluorescence signals. This seems to suggest confocal microscopy for optical imaging, but some constraints prohibit its straightforward application for the detection of intrinsic optical signals. Firstly, a fluorescing object as usually detected by confocal microscopes, acts as a light source, and the origin of the emitted photons and therefore the location of the object in depth can be well detected by confocal optics. In contrast, absorption and scattering signals (including intrinsic signals) are sampled over the whole volume passed by an ensemble of photons and therefore information about the depth of local absorbers or scatterers cannot be obtained in an obvious way. Secondly, due to technical reasons an optical imaging setup requires a recording technique which can keep some minimum working distance of a few centimeters from the brain surface. Because confocal microscopes work in the regime of high numerical apertures, their maximum distance from the specimen lies far below the required working distance. Finally, intrinsic signals are very weak and due to signal-to-noise reasons we have to detect many photons per voxel (cf section 1). Due to the strong restriction imposed by the confocal pinhole, confocal

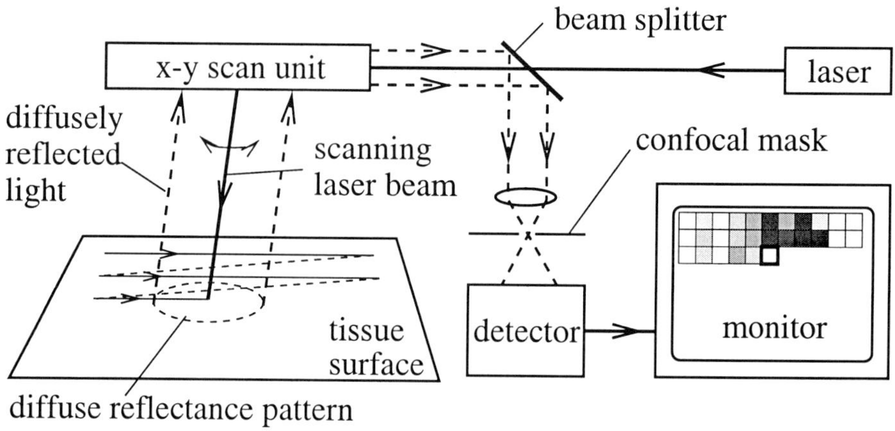


*Figure 4.12.* Basic principle of scanning laser imaging. A thin laser beam scans line by line over the surface of the tissue. For each position of the scanning laser beam, the diffusely reflected light is collected, unscanned, and sent, eventually through a confocal mask, onto a light detector. The amount of detected light is used to determine the gray level of a single corresponding pixel. The total image is generated pixel by pixel and line by line for all positions of the scanning laser beam.

microscopes usually detect only a few tens of photons per voxel, a number far too small for optical imaging.

In this section we suggest an experimental setup for high contrast depth selective optical imaging and, again using Monte Carlo simulations, characterize, to what extent this technique should outperform conventional video optical imaging. Similarly to confocal microscopy, the basis for the proposed setup is formed by a scanning laser device, but here we suggest to run the system in a weakly or non-confocal ("flying spot") mode. This provides the possibility of a large working distance and a high photon yield.

## 4.1 Principle and Simulation of Scanning Laser Imaging

Figure 4.12 schematically illustrates the basic principle of scanning laser imaging. In this technique, a thin, collimated laser beam scans line by line over the object under consideration, which in our case is a part of the tissue surface. For a given position of the beam, the diffusely reflected light is unscanned by propagating it through the x-y-scan unit in the reverse direction, and is guided (eventually through a confocal mask) to a light detector. This reflected light is used to determine the gray level of a single pixel within the image frame, the position of which within the frame is determined by the current orientation of the scanning laser beam and thus corresponds to the currently illuminated small patch of cortical tissue. During one complete scan, the gray levels of all pixels within the image frame are measured in sequence. In particular, because the whole distribution of the diffusely reflected light – the diffuse reflectance pattern – is in principle available for every single pixel, one can hope to be able to detect more subtle properties of intrinsic signals (such as their 3D-positions) from the details of the reflectance patterns.

For the simulation of scanning laser optical imaging, we calculate the diffuse reflectance patterns for a two-dimensional grid of photon injection sites, identically to video optical imaging. Simulated scanning laser images are obtained by calculating a given property of the diffuse reflectance pattern for each pixel separately and attaching the corresponding value to that pixel. For example, if only intensity information is to be recorded, the total remitted photon count is determined for each of the $(2M + 1)(2M + 1)$ reflection patterns and is assigned to the single raster point, around which the corresponding reflection pattern is centered.

Actually, the analysis presented in the following sections is based on the very same set of reflectance patterns as used for simulated video imaging in section 3. Hence, the differences in the images derived from both methods result from different kinds of the light detection rather than from different kinds of illumination or the detection of different tissue properties, since the identical photon propagation pathways are used for both video and scanning laser optical imaging. This enables us to provide a reliable comparison of the relative changes in image resolution and contrast.

## 4.2 Lateral Resolution and Contrast

In this subsection we perform a comparison between simulated scanning laser images and video images for the same model tissue as described in section 3.1: $N = 10^6$ photons were injected into each of $21 \times 21$ illumination points and the diffuse reflectance patterns were calculated. Scanning laser optical imaging was simulated by summing over the total reflected light for each photon injection $(x, y)$ separately and taking the resulting value as the gray level $I(x, y)$ of a single pixel corresponding to the photon injection site. No confocal geometry was used for these simulations corresponding to a "flying spot" geometry.

Figure 4.13 compares the resulting scanning laser images (left column) with the video images (right column and figure 4.10) for absorber depths of $Z = 0.2$ and 0.6 mm respectively. The scanning laser images are much less blurred and stronger in contrast than video images. As a consequence, which becomes obvious from a comparison of the two bottom plots, we predict that the proposed scanning laser method can image deeper structures than the video technique. The image contrast, eq. (4.20), has been evaluated for scanning laser images and is compared to the corresponding contrast reached by conventional video imaging in figure 4.14. For all depths, the contrast of the scanning laser image is approximately threefold higher than that of the corresponding video image.

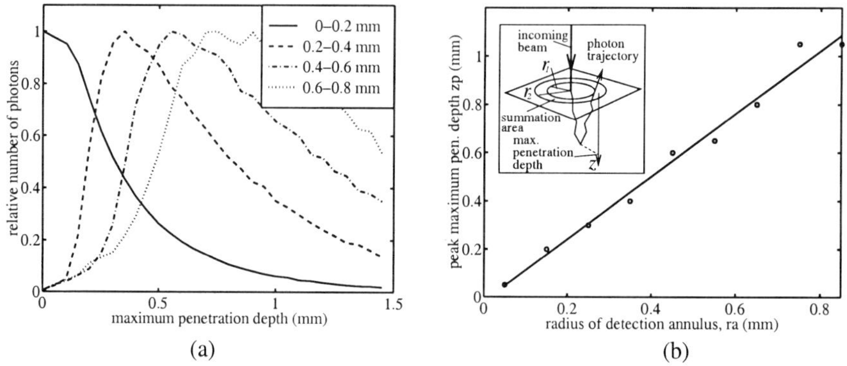


*Figure 4.15.* **(a)** Normalized penetration depth histograms for $10^8$ photons injected into a homogeneous medium, averaged over annuli with inner radii $r_1 = 0, 0.2, 0.4, 0.6$ mm and outer radii $r_2 = r_1 + 0.2$ mm (see inset in **(b)**). The most frequent penetration depth increases systematically with the radius of the annulus. Maximum photon counts: $0 - 0.2$mm: 77900, $0.2 - 0.4$ mm: 25014, $0.4 - 0.6$ mm: 12285, $0.6 - 0.8$ mm: 7472. **(b)** The peak penetration depth depends approximately linearly on the radius $r_a = (r_1 + r_2)/2$ of the detection annulus. Solid: linear least squares fit, offset -0.02 mm, slope 1.30, circles: data. Inset: Calculation of the penetration depth histograms vs. the lateral spread of the photons. The histogram is built from all photons that were exiting the medium within an annulus between $r_1$ and $r_2$ (detection annulus). For these photons, the different possible maximum penetration depths were counted in order to determine the penetration depth histogram.

select the detected photons by their different penetration depths. Figure 4.15b plots the most frequent penetration depths $z_p$ of the photons (i.e. the maxima of the functions in figure 4.15a) vs. the radius $r_a$ of the detection annulus and reveals a linear relationship between both quantities. The least squares fit of a linear function yields

$$z_p = 1.30 r_a - 0.02. \tag{4.23}$$

Selective measurement of photons that exit from an annulus around the illuminated origin can be performed using standard confocal geometry by application of an annulus-type confocal mask, which is optically conjugate to the surface of the tissue (cf. figure 4.18). Equation (4.23) establishes a first order relationship between the radius of the detection annulus and the depth up to which intrinsic signals are sampled. At the same time it becomes obvious that the penetration depth histograms have a considerable spread around their peeks, and a clear separation of photons by penetration depth is not possible.

Now we compare the diffuse reflectance patterns for a single incident light beam obtained from a homogeneous tissue (see previous subsection) with that obtained from the same tissue equipped with a local absorber at $\mathbf{R} =$

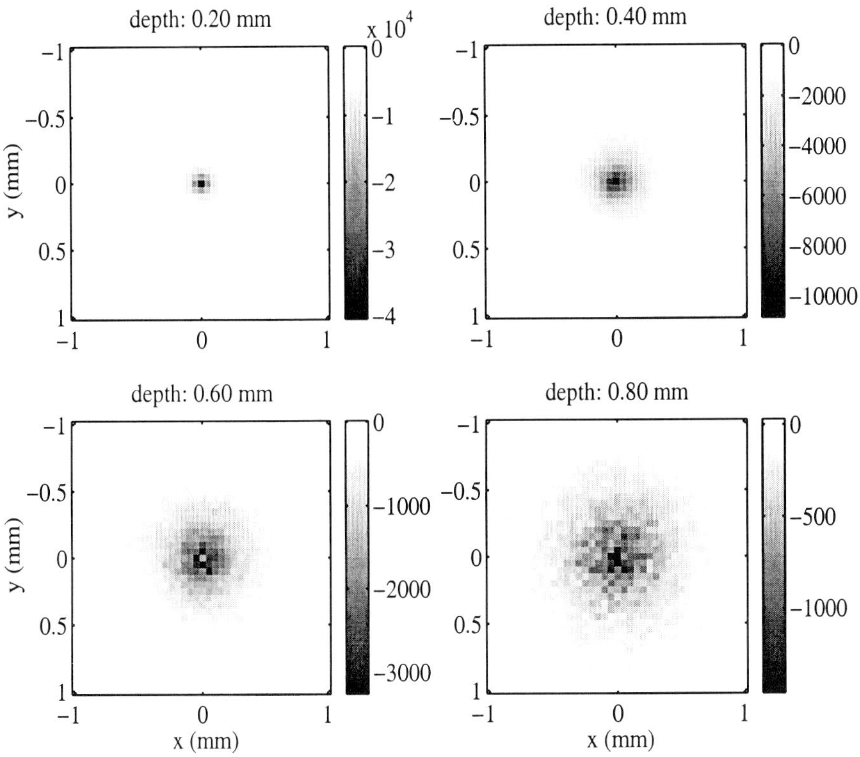


*Figure 4.16.* Absorption patterns, which are defined as the differences between the reflectance patterns calculated with and without an absorber being present at depths $Z = 0.2, 0.4, 0.6, 0.8$ mm below the photon injection point. Gray levels indicate the difference in photon count per raster square. With increasing depth of the absorber, the signal amplitude strongly decreases, while the fraction of darkening at the periphery of the absorption patterns (far from the origin) increases.

$(0, 0, -Z)$ (i.e. below the the photon injection point). Diffuse reflectance patterns with $10^8$ injected photons each were calculated for vertical positions $Z = 0.2, 0.4, 0.6, 0.8$ mm of the local absorber. The difference between the two-dimensional diffuse reflectance patterns of the homogeneous medium and the medium plus an absorber represents the changes in reflected light that are are caused by the absorber.

Figure 4.16 displays these difference patterns for the four depths of the absorber on a $41 \times 41$ raster with 0.05 mm grid constant. The extent of darkening represents the decrease of reflected light intensity for each raster point due to the absorbing sphere. The result shows, that the strength of the absorption signal strongly decreases with increasing depth of the absorber. This is in agreement with the results presented in figure 4.13. Further, besides the amplitude of the

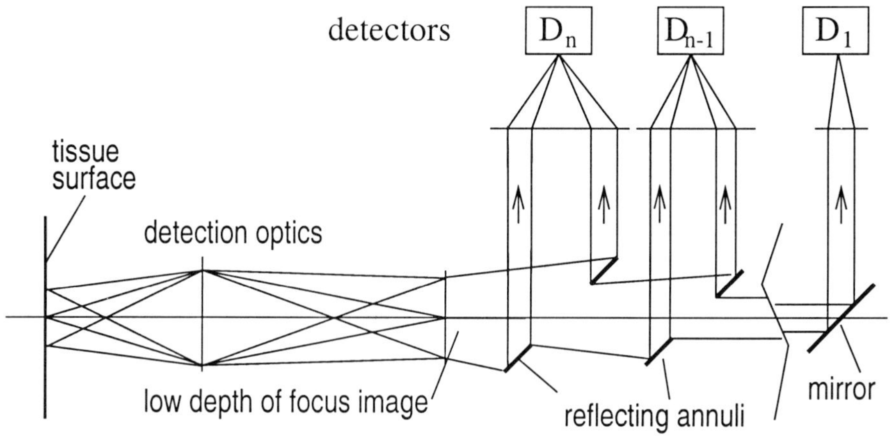


*Figure 4.18.* Schematic illustration of online-detection of the radial reflectance function. An optical setup forms a low depth of focus image of the diffuse reflectance pattern. This image propagates through a set of reflecting annuli with systematically decreasing radii. Each annulus reflects a ring-shaped part of the diffusely reflected light onto one of $n$ detectors $D_1, ..., D_n$ which thereby detect light form a single annulus within the reflectance pattern.

Currently available scanning laser devices such as scanning laser ophthalmoscopes (SLO) use avalanche diodes for light detection (Tornow et al., 1997), which have single photon sensitivity and very short readout times. SLO's can provide images in video resolution (equivalent to 768 $\times$ 512 pixels) in video frame rate (Webb and Hughes, 1981; Fitzke et al., 1991; Fitzke and Masters, 1993; Stetter et al., 1996; Stetter et al., 1995d; Stetter et al., 1995e). A second suggestion for a possible implementation of scanning laser imaging is illustrated schematically in figure 4.18. The detector unit of a SLO can be replaced by a cascade of ring-shaped mirrors, the apertures of which become smaller in diameter with increasing distance from the reflected light to be detected. A confocal optical setup with a high depth of focus generates an image of the diffuse reflectance pattern on the first mirror. This mirror reflects the most peripheral part of the reflectance pattern to a detector $D_n$ and passes the remaining light to the second mirror, which in turn reflects the remaining most peripheral part of the light to detector $D_{n-1}$ etc., until the central part of the incoming light hits the last detector $D_1$. The $n$ detectors provide the radial reflectance function $R(r_a)$ in real time, i.e. with the full temporal and intensity resolution of the detectors.

The 3D-distribution of changes in light absorbance could then be estimated as follows: First, a full 2D-set of radial reflectance functions, i.e. one function for each pixel of the frame, is measured using the proposed setup. The simulations show, that the different detection annuli of that setup to some extent select photons by their penetration depths into the tissue (see figure 4.15a). In addition, the most frequent penetration depth depends roughly linearly on the radius of the considered detection annulus (figure 4.15b). The photons detected by a given annulus can be assigned a quantitative penetration depth via this relationship. If the penetration depth selectivity by each annulus were perfect, the changes in the radial profile for one pixel would be just the cumulative change in absorption (its integral from the surface up to the depth under consideration), as soon as the abszissa is scaled to reflect the photon penetration depths via the relationship mentioned above. Thus the first derivative of the reflectance function of a given pixel would yield the vertical distribution of absorbance changes below that pixel. However, because the penetration depth selectivity is not perfect but consists of rather broad histograms (figure 4.15a), the 3D-distribution obtained from this procedure represents a convolution of the original distribution with an unknown convolution kernel. 3D blind deconvolution techniques, which have been recently developed for confocal microscopy, might help to recover the original model distribution from a set of simulated radial reflectance profiles. These deconvolution techniques could be guided by some prior knowledge about the convolution kernels obtained from the simulations presented here.

## 4.5 Influence of Variations in the Simulation Setup

In this subsection we discuss to what extent different assumptions made for the simulation setup could influence the predictions obtained. All tissue optics simulations were carried out using a single set of optical parameters taken from human gray matter in vitro, but what happens if these numbers are not completely correct? First of all, though the true values and their variances to be expected in vital cortical tissue are not known, all optical parameters of biological tissues are relatively similar to each other (Cheong et al., 1990). this suggest, that the true parameters do not deviate very strongly from the ones taken in the presented simulations. It should be kept in mind, however, that our optical parameters were obtained on blood-free preparations and blood has a considerably higher scattering and absorption coefficient than the brain tissue, and the in vivo values for $\mu_a$ and $\mu_s$ could be higher than those taken here.

Variations in the absorption coefficients leave the trajectories of the photons unchanged, but lead to the absorption of a different fraction of them. As can be seen from eq. (4.19), if the absorption coefficient increases from $\mu_a$ to $\mu_a + \Delta\mu_a$, the survival probability for the homogeneous medium changes to $P_a(l) = \exp(-\mu_a l)\exp(-\Delta\mu_a l)$, such that photons with longer pathways through the tissue are more strongly absorbed than those with short trajectories. Hence, an increase in $\mu_a$ would cause a stronger lateral decay of the diffuse reflectance pattern. Similarly, a decrease in the forward scattering anisotropy $g$ or an increase in the scattering coefficient $\mu_s$ would similarly cause a lateral shrinkage of the diffuse reflectance pattern eq. (4.11). In contrast, changes in the incremental absorption coefficient $\mu_A$ of the local absorber does not change any of the results qualitatively, since for a given set of photon trajectories, the number of photons that hit the absorber remains constant. Therefore, the only effect of an increase of the absorption coefficient by $\Delta\mu_A$ is a decrease the fraction of photons, which can actually pass the absorber by a factor in the order of $\exp(-2\Delta\mu_A\rho)$, where $\rho$ is the absorber radius. Due to this reason, we strongly increased the absorption coefficient of the local absorber compared to a biologically realistic value, because this reduces the number of injected photons necessary to reveal its effect. If one aims at the investigation of the images of several absorbers or interacting absorbers and local scatterers, it may become important to take more realistic values for the absorption coefficient $\mu_A$. For example, the detection of two absorbers, one being located above the other, may occur roughly independent of each other because of the small size of the change (0.1 % change of the backscattered light). However, if both of them are provided with very strong absorption coefficients in a simulation, the deeper absorber may become even undetectable because the more superficial absorber removes all the photons attempting to pass it. This may in fact occur in the case of superficial cortical blood vessels, which weaken both the incoming and the reflected light and thereby probably influence the strengths of the detected intrinsic signals (McLoughlin and Blasdel, 1997).

For the simulations, we assumed a vertical illumination of the model tissue, which is only approximately fulfilled for scanning laser techniques and is practically not fulfilled for conventional video imaging. Instead, the cortex is illuminated laterally under a relatively small angle in order for the video camera to have access to the cortical surface. This kind of illumination should further decrease the penetration depth of the light and increase its lateral spread thus further decreasing the image quality reached by the video imaging technique. This points to a further advantage of scanning laser detection, where illumination and light detection occur along the same optical pathway thus allowing an approximately vertical illumination in a natural way.

Finally, all simulations assume a geometrically flat and optically isotropic tissue, whereas biological tissue is definitely non-isotropic and curved. The curvedness of realistic tissues may not cause strong problems for scanning laser imaging, since the proposed light detection method involves a high depth of focus. However, possible influences of tissue anisotropies have to be tested both by simulations and by experiments.

Possible experimental tests of scanning laser imaging could be performed by first injecting small local absorbers such as microbeads or small patches of ink into nerve tissue, then imaging the tissue and afterwards histologically determining the location and extension of the absorbers. If these experimental tests confirm our simulations, we may have the key to a new generation of devices for the detection of three-dimensional neural activity distributions with high contrast, high resolution, and relatively low-cost compared to leading-edge fMRI techniques (Kim et al., 2000a).

# Chapter 5

# OPTICAL IMAGING AS SOURCE SEPARATION PROBLEM

If we want to reliably infer neuronal activity patterns from optical imaging data, we face two problems: ($i$) Most of the optical signals represent only an indirect measure of neuronal activity. This is particularly true for evoked metabolic changes such as intrinsic signals. Therefore, the relationship between neuronal activity and optical signals must be characterized in detail. The only direct method for doing this is the simultaneous electrophysiological and optical measurement of neuronal activity, which has been done for intrinsic signal imaging (Malonek and Grinvald, 1996). We will not further treat direct comparison, but will later discuss some indirect methods for estimating to what extent intrinsic signal components reflect neuronal activity. ($ii$) Only some of the measured changes in light reflectance are suitable for monitoring neuronal activity. They are mixed with unwanted components, which either reflect only the global activation level or are independent of neuronal activity. We need to separate the signal components in the image stack, which are closely and locally coupled to neuronal activity, from the remaining components. In other words, dye signals or intrinsic mapping signals must be reliably separated from global signals, blood vessel patterns, hemodynamics and ongoing activity in order to obtain a (still indirect) measure for two-dimensional neuronal activity distributions over space.

This and the following chapters 6 – 8, summarize recent approaches towards the solution of problem ($ii$): "extract the relevant optical signals from raw data". Here we first formulate the analysis task for optical imaging data as a source separation problem. Then we describe some of the standard methods for solving this source separation task, but also list a few of their limitations, which motivate the development and application of alternative analysis methods presented later.

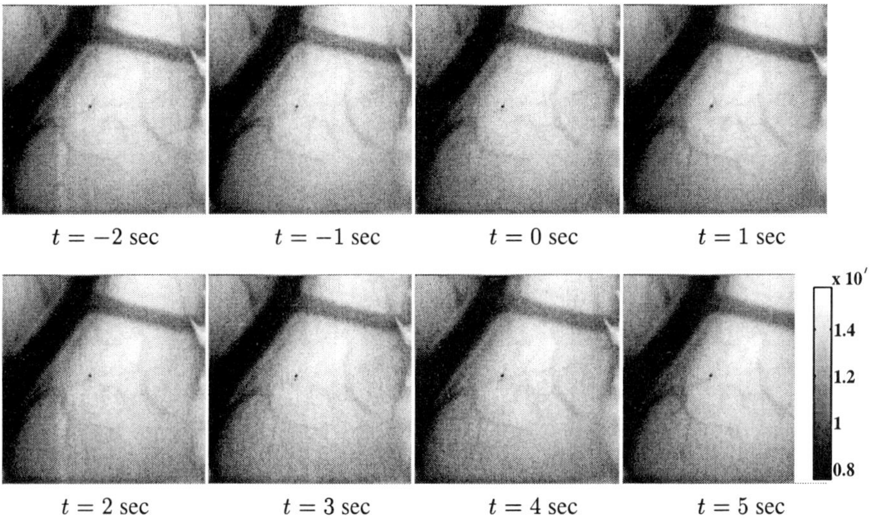


*Figure 5.1.* Summed and binned raw data set from optical imaging of intrinsic signals. Images show a 3.85 × 3.85 mm region of V1 of a macaque monkey. Each frame contains 1 second of data collection. Subscripts denote times relative to the stimulus onset, at which the measurement of the corresponding frame started. Reflectance changes due to intrinsic signals are much too small to be visible.

# 1. Formulation of the Source Separation Task

During an optical imaging trial, a CCD-camera collects a set of images from the illuminated brain surface before, during, and after stimulation of the corresponding nerve tissue. Let us first specify a mathematical notation for the description of an optically recorded image stack.

## 1.1 Data Representation

If we illuminate a cortical region, the intensity of the light reflected at time $t$ from position $\mathbf{r}$ of the cortical surface is denoted as $x(\mathbf{r}, t)$. During an optical imaging experiment, the reflected light $x$ is projected by the camera lens onto the photo diode array of the CCD chip. The photo diode array of the camera contains a raster of $P_x \times P_y$ pixels, and the camera samples the intensity distribution every $\Delta t$ seconds to yield $M$ frames $m = 1, ..., M$, with $P = P_x \times P_y$ pixels each. For small pixels (each corresponding to a square $\Delta r$ mm in size on the cortex) and for short time intervals $\Delta t$, the number of photons collected at pixel $\mathbf{r} = (r_x, r_y)$, $1 \leq r_x \leq P_x, 1 \leq r_y \leq P_y$ between time $t_m$ and $t_m + \Delta t_m$ is approximately given by

$$x_m(\mathbf{r}) = \eta x(\mathbf{r}, t_m) \Delta t \Delta r^2, \tag{5.1}$$

where $\eta$ is a constant proportionality factor, which depends on device parameters including the quantum efficiency of the CCD chip and $m$ is the frame number. If we assume a constant geometry of the setup and a constant frame collection rate, $\Delta r$ and $\Delta t$ can be absorbed into the proportionality factor, which we henceforth omit for simplicity.

Figure 5.1 shows an example of an image stack, as it is obtained from an optical imaging experiment. For this particular data set, images with $256 \times 256$ pixels – corresponding to a $3.85 \times 3.85$ mm region – of the primary visual cortex of a macaque monkey were taken under illumination with 605 nm and with a frame rate of 15 Hz (camera: SMD–1M60). The gray-level of each raw frame is described by a 16 bit integer. For every single trial, a sequence of 120 frames was taken during 8 seconds, while from $t = 0$ s up to $t = 5$ s one eye was stimulated by a drifting square wave grating. The data collection sequence was followed by a 16 s recovery period without visual stimulation, after which the next trial started. Sixteen identical trials were summed up pixelwise to yield a single averaged image stack, and sequences of 15 subsequent frames of this stack were binned to yield the 8 frames $x^0_m(\mathbf{r}), m = 1, ..., 8$ shown in figure 5.1. Summation and binning was carried out in order to reduce photon shot noise by increasing the effective number of photons contributing to each pixel (cf. chapter 4, section 1.1).

### 1.1.1 Preprocessing of Data

The images in figure 5.1 show a large and several smaller superficial blood-vessels, but there are no obvious stimulus-evoked changes observable. This is the case because the intrinsic signals are very small compared to the background reflectance. The small changes can be emphasized by subtraction of the first frame from the other frames:

$$x'_m(\mathbf{r}) = x^0_m(\mathbf{r}) - x^0_1(\mathbf{r}). \tag{5.2}$$

This operation removes the time-independent components of the light reflectance and is referred to as first frame analysis (Bonhoeffer and Grinvald, 1996).

Figure 5.2 displays the same data set as figure 5.1 after first frame analysis. Now it is possible to observe a variety of changes in reflectance with different spatial patterns: One prominent signal consists of an increase in reflectance from the larger blood vessels. Also, some smaller vascular patterns are visible, which are caused by small movements of the blood vessels relative to the brain surface. Further, the whole parenchyma becomes darker in the late phase of the trial. This effect can be categorized as a global signal (cf. figure 4.3) and is due to an unspecific increase in parenchymal blood-volume. Finally, there is a barely visible system of three approximately horizontal bands, which represent

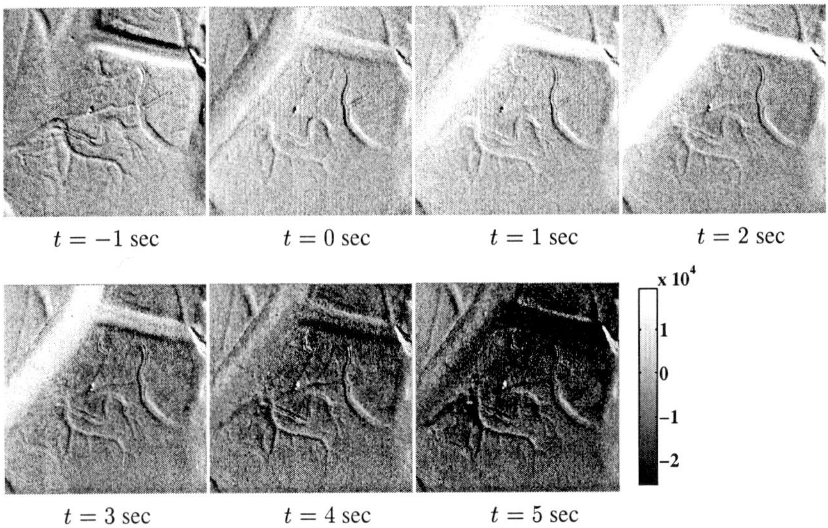


*Figure 5.2.* Same single condition stack as shown in figure 5.1, but after first frame analysis. The first zero frame is not shown anymore. The most prominent changes in time are an increase in reflectance in the larger blood vessels and a darkening due to a blood-volume related global signal in the late phase. Changes range around 0.1 percent of the background reflectance (cf. figure 5.1).

the mapping signal of this experiment. A comparison between the scale bars in figure 5.2 and figure 5.1 demonstrate, that all these changes in reflectance amount to approximately 0.1 percent of the background reflectance.

Figure 5.3 shows every second frame of figure 5.2 again, but now the big blood-vessel has been masked out and an individual gray scale has been applied to each image in order to emphasize the mapping signal. The mapping signal consists of a set of roughly horizontal bands, which we can identify as the ocular dominance bands of the left-dominated cells. The figures demonstrate that the mapping signal is usually strongly contaminated by other signal components, which can be much stronger in amplitude than itself. If we want to use the mapping signal as an estimator of neuronal population activity, we have to find data processing methods, which allow us to reliably separate it from the other signal components based on the measured mixtures in figures 5.2 and 5.3.

## 1.2 The Source Separation Problem and Data Model

Let us reformulate the problem we want to solve:

- In an optical imaging experiment, we measure a mixture of different signal components, which are related to different types of intrinsic signals.

$t = -1$ sec $\quad t = 1$ sec $\quad t = 3$ sec $\quad t = 5$ sec

*Figure 5.3.* Every second image of the same data set as in figure 5.2, but after masking out the big blood vessels and with individual gray scales for better visibility of the mapping signal. For each image, the gray level was set to the mean $\pm$ 1.8 times the standard deviation of gray levels for the masked region.

- From these measurement we want to extract one signal component, namely the mapping signal, i.e. we wish to separate it from the other components.

- We do not know the patterns of individual signals nor do we have information about the mixing process, e.g. the mixing coefficients of a linear mixture.

- Finally, the mixtures are strongly contaminated by photon shot noise.

In the signal processing community, this problem is well-known as Blind Separation of Sources (BSS) problem. For optical imaging, the sources correspond to the individual signal components such as the mapping signal, global signal, vascular patterns etc. We want to separate the sources from each other on the basis of our data sets which only contain mixtures of sources. The adjective "blind" denotes that the separation is attempted without any explicit knowledge about the sources or the mixing process but is based on assumptions about the statistical nature of the sources. The BSS task has to be carried out in the presence of noise, hence optical imaging rises a noisy BSS task.

We formulate a statistical model for the collected data set as follows: The mean reflected light is assumed to consist of several components, namely the mean background reflectance $R_0(\mathbf{r})$ and changes in reflectance $R_l(\mathbf{r}, t)$, which are caused by different intrinsic signals indexed by $l = 1, ..., L$. $R_l(\mathbf{r}, t)$ is the $l$-th cause or the $l$-th signal source function. For example, $R_1(\mathbf{r}, t)$ could describe the spatiotemporal course of the mapping signal, $R_2(\mathbf{r}, t)$ of the global signal, $R_3(\mathbf{r}, t)$ spatiotemporal vascular patterns, etc. Because the sources are small compared to the background, it seems reasonable to neglect higher order terms, because multiplicative terms of the form $R_j(\mathbf{r}, t)R_l(\mathbf{r}, t)$, which describe interactions between sources, are much smaller than linear terms in the sources. Motivated by this consideration, we formulate the following linear statistical model:

$$x_m^0(\mathbf{r}) = R_0(\mathbf{r}) + \sum_{l=1}^{L} R_{m,l}(\mathbf{r}) + n_m(\mathbf{r}). \tag{5.3}$$

In eq. (5.3), $R_{m,l}(\mathbf{r}) = R_l(\mathbf{r}, t_m)$ denotes a sample of the $l$-th source at time $t_m$, and the random variables $n_m(\mathbf{r})$ describe photon shot noise. In its most general form, the BSS task consists of the recovery of the mapping signal, say $R_{m,1}(\mathbf{r})$, from the measured mixture $x_m^0(\mathbf{r})$.

The random variables $n_m(\mathbf{r})$ in eq. (5.3) describe the fluctuation of the actual photon count around its mean value (photon shot noise). Photon shot noise at different times and within different pixels is assumed statistically independent and Poisson-distributed. Because in optical imaging the mean photon count is usually strong (cf. section 1.1), the corresponding Poisson distribution can be well approximated as a Gaussian distribution with a variance proportional to the total mean reflectance $R(\mathbf{r}, t_m)$. The probability distribution function for photon shot noise becomes

$$p(n_m(\mathbf{r})) = \frac{1}{\sqrt{2\pi}\sigma_m(\mathbf{r})} \exp\left(-\frac{n_m(\mathbf{r})^2}{2\sigma_m^2(\mathbf{r})}\right), \quad \sigma_m^2(\mathbf{r}) = \sigma_0^2 R(\mathbf{r}, t_m), \tag{5.4}$$

where $\sigma_0$ denotes a constant factor, which is related to camera properties including the quantum yield. If we take into account the fact, that the total mean reflectance deviates only very little from the value of the background reflectance, we can approximate the noise variance by

$$\sigma_m^2(\mathbf{r}) \approx \sigma_0^2 R_0(\mathbf{r}) \equiv \sigma_n^2(\mathbf{r}), \tag{5.5}$$

i.e., the noise variance may depend on space, but is almost independent of time.

In the present context, the background reflectance does not contain any information about the signal sources, but constitutes most of the reflected light. Therefore, we will only consider data sets, from which the background reflectance has been removed by first frame analysis. The model becomes:

$$x_m(\mathbf{r}) = \sum_{l=1}^{L} R_{m,l}(\mathbf{r}) + n_m(\mathbf{r}), \tag{5.6}$$

or in matrix formulation

$$\mathbf{X} = \sum_{l=1}^{L} \mathbf{R}_l + \mathbf{N}. \tag{5.7}$$

In eq. (5.7), all matrices have the size $M \times P$. The components of the data matrix $\mathbf{X}$ are $(\mathbf{X})_{m,\mathbf{r}} = x_m(\mathbf{r})$, and likewise the $(m, \mathbf{r})$-th component of $\mathbf{R}_l$ and $\mathbf{N}$ are $R_{m,l}(\mathbf{r})$ and $n_m(\mathbf{r})$, respectively. Column vectors of the data matrix are the time series of gray values for an individual pixel $\mathbf{r}$ and are denoted by $\mathbf{x}(\mathbf{r}) = (x_1(\mathbf{r}), ..., x_M(\mathbf{r}))^T$ (the subscript $T$ denotes the transpose of a matrix or vector). They are referred to as pixel time series. The rows of the data matrix contain the pixels of the $m$-th image of the image stack $\tilde{\mathbf{x}}_m = (x_m(1,1), ..., x_m(P_x, P_y))$. The same scheme holds for the other matrices.

Blind source separation cannot be performed without any prior knowledge or prior assumptions about the behavior of the sources. There are a variety of more or less reasonable assumptions about the sources, which can serve as a basis for a solution of the source separation task. They include the following statements:

- The sources occupy different bands in the frequency domain.
- The sources can be distinguished by their variance over the data set.
- Different sources are uncorrelated or statistically independent from each other.
- The sources change smoothly over space and time.

## 1.3 Spatiotemporally Separable Problems

While various sources of knowledge can be used for solving the BSS task, the separation of $L$ small sources from a single noisy measurement is still a hard problem. Therefore, most of the analysis methods treated in this chapter make use of a further assumption, which strongly reduces the number of possible patterns of the sources: separability of the sources in space and time. This assumption may not be strictly fulfilled, but seems not unrealistic in light of the observed data. From figure 5.2 it can be seen that the different sources do not strongly change their spatial patterns, but only change their amplitudes over time. The reflectance changes in blood-vessels are spatially restricted to the morphological vascular pattern. Also, the global and mapping signals occur where neuronal activity is evoked. Even if the spatial pattern of neuronal activity would change over time, this change would probably be too fast for the intrinsic signals to follow. It seems therefore reasonable to assume that the mapping and global signals do not strongly change their spatial patterns as well, which seems to agree with the data (figure 5.3). Finally, even small movement artifacts such as the patterns visible at some smaller blood vessels in figure 5.2 can be described by a single spatial pattern, namely the gradient of the blood-vessel profile.

Figure 5.4 summarizes this view of an optical imaging data stack: The spatiotemporal signal can be described by a set of spatial prototype patterns which pop up and vanish again in response to stimulation. Consequently, the signal of each frame can be described as a linear combination of the same set of spatial prototype patterns, yet with different mixing coefficients. In mathematical

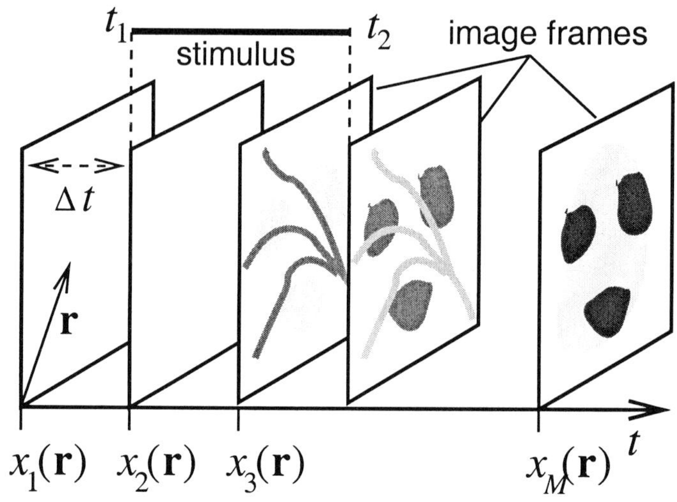


*Figure 5.4.* Illustration for spatiotemporal separability in optical imaging experiments. According to that assumption, a set of constant spatial prototype patterns pops up and vanishes again with different time courses, and each frame contains a different linear combination of the prototype patterns.

terms, each source $R_l(\mathbf{r}, t)$ can be written as

$$R_l(\mathbf{r}, t) = a_l(t) s_l(\mathbf{r}), \tag{5.8}$$

and the data model becomes

$$x_m(\mathbf{r}) = \sum_{l=1}^{L} a_{m,l}\, s_l(\mathbf{r}) + n_m(\mathbf{r}), \tag{5.9}$$

where the mixing coefficients are related to the time courses of the respective source pattern via $a_{m,l} = a_l(t_m)$. Finally, if we combine the mixing coefficients to the $M \times L$ mixing matrix $\mathbf{A}$, $(\mathbf{A})_{m,l} = a_{m,l}$ and the sources to the source vector $\mathbf{s}(\mathbf{r}) = (s_1(\mathbf{r}), ..., s_L(\mathbf{r}))^T$, we can formulate the data model for linear Blind Source Separation as

$$\mathbf{x}(\mathbf{r}) = \mathbf{A}\ \mathbf{s}(\mathbf{r}) + \mathbf{n}(\mathbf{r}). \tag{5.10}$$

This model describes $P$ observed pixel time series as realizations of a generative model, which produces data vectors as noisy mixtures of $L$ sources. Neither the sources $\mathbf{s}$ nor the mixing matrix $\mathbf{A}$ are known. The BSS problem consists of estimating both $\mathbf{A}$ and $\mathbf{s}$ in eq. (5.10) from the observed data set $\{\mathbf{x}(\mathbf{r})\}$ by use of some general assumption about the sources or source statistics. The following section summarizes approaches towards the solution of the BSS problem eq. (5.10) or eq. (5.6), which have been established in the optical imaging literature.

# 2. Heuristic Methods for Source Separation

Three widely used procedures for post-processing of optical imaging data stacks are differential imaging (Blasdel, 1992a), subtraction of or division by a "cocktail blank" and bandpass filtering (Bonhoeffer and Grinvald, 1996). Often, bandpass filtering and one of the other methods are combined for the extraction of the mapping signal. In this section we provide a description of these methods in the framework of our data model for optical imaging and specify the assumptions which they use for solving the BSS problem (5.9), but also list a few limitations which motivate alternative approaches treated in chapters 6 and 7.

## 2.1 Differential Imaging

If an optical imaging data set is based on a single type of stimulus (denoted as one stimulus condition), as shown in figure 5.2, it is called a single-condition image stack. For some aspects of cortical function there exist pairs of stimuli, which are already known to excite disjunct populations of neurons. A pair of stimuli which fulfills this condition is called an orthogonal pair. Examples for approximately orthogonal stimuli are left-eye and right eye stimulation (cell responses are approximately disjunct because of ocular dominance) and oriented stimuli which differ by 90 deg in orientation (because of orientation selectivity of most cortical neurons, cf. figure 3.11). If disjunct neuron populations are activated by orthogonal stimuli, we expect the mapping signal for both stimulations also to cover disjunct areas. In contrast, global changes in light reflectance only respond to the mean activation over a larger cortical region, which is similar for both stimuli.

Differential imaging is based on the model assumption that all signals except the mapping signal are identical for orthogonal stimuli, whereas the two mapping signals are complementary to each other. If this model is true, and denoting the mapping signal as the first source, $s_1(\mathbf{r})$, the responses to the pair of orthogonal stimuli 1 and 2 can be written as

$$x_m^{(1)}(\mathbf{r}) = \sum_{l=2}^{L} a_{m,l}\, s_l(\mathbf{r}) + a_{m,1}\, s_1(\mathbf{r}) + n_m^{(1)}(\mathbf{r}), \tag{5.11}$$

$$x_m^{(2)}(\mathbf{r}) = \sum_{l=2}^{L} a_{m,l}\, s_l(\mathbf{r}) + a_{m,1}\, (s_{\max} - s_1(\mathbf{r})) + n_m^{(2)}(\mathbf{r}), \tag{5.12}$$

where the superscripts denote the stimulus condition and $s_{\max}$ denotes the maximum value of the mapping signal. We can estimate the mapping signal by subtracting the two image stacks from each other,

$$\Delta x_m(\mathbf{r}) = \frac{1}{2}(x_m^{(2)}(\mathbf{r}) - x_m^{(1)}(\mathbf{r})) \tag{5.13}$$

$$= \frac{1}{2}\left(a_{m,1}\left(2s_1(\mathbf{r}) - s_{\max}\right) + n_m^{(1)}(\mathbf{r}) - n_m^{(2)}(\mathbf{r})\right) \quad (5.14)$$

and averaging the result over time yields:

$$\hat{s}_1(\mathbf{r}) = \frac{1}{M}\sum_m a_{m,1}\left(s_1(\mathbf{r}) - \frac{s_{\max}}{2}\right) + \frac{1}{M}\sum_m \frac{1}{2}\left(n_m^{(1)}(\mathbf{r}) - n_m^{(2)}(\mathbf{r})\right) \quad (5.15)$$

$$= \bar{a}_1 s_1(\mathbf{r}) - \bar{a}_1 \frac{s_{\max}}{2} + n'(\mathbf{r}). \quad (5.16)$$

The hat in eq. (5.16) marks the estimate of a statistical quantity and $\bar{a}_1$ abbreviates the mean over the time series $a_{m,1}$. The mapping signal can be correctly estimated up to a constant factor, and if the noise is uncorrelated, the standard deviation of the averaged noise $n'(\mathbf{r})$ is reduced by a factor $\sqrt{M}$: $\sigma_{n,\mathrm{df}}(\mathbf{r}) = \sigma_n(\mathbf{r})/\sqrt{M}$.

Figure 5.5 demonstrates the procedure of differential imaging and uses the same example as shown in figures 5.1 - 5.3. Single condition image stacks for left-eye and right eye stimulation are shown in the top and middle row of figure 5.5a, and the bottom row contains their pixelwise difference. Blood vessel patterns and the global components are strongly reduced, whereas the mapping signal is preserved. The resulting differential image has been obtained as the mean over the last four images of the differential image stack and is shown in figure 5.5b. The approximately horizontal ocular dominance bands pop up in the picture.

### 2.1.1 Limitations

There are two major limitations for the applicability of differential imaging. In summary: $(i)$ Global signals may not always be identical for different conditions and $(ii)$ the design of orthogonal pairs of stimuli requires knowledge about cortical organization, and orthogonal pairs also do not always exist for arbitrary experiments. The assumption that all signals except the mapping signal are identical in two data sets for orthogonal stimuli represents an idealized situation. Differences can arise from several origins: $(i)$ Variability in the response of global stimulus-locked components. This would lead to an incomplete removal of the global components from the differential image stack. Because global signals usually are stronger than the mapping signal, even a small residual of a global signal would mean a strong contamination of the mapping signal. Particular examples for signal variability are small movement artifacts of superficial blood vessels (figure 5.5a,bottom,leftmost image) or variability in the strength of blood-flow and blood-volume changes (visible as bright blood-vessel pattern in figure 5.5a,bottom,images 5-7 from left). $(ii)$ Different

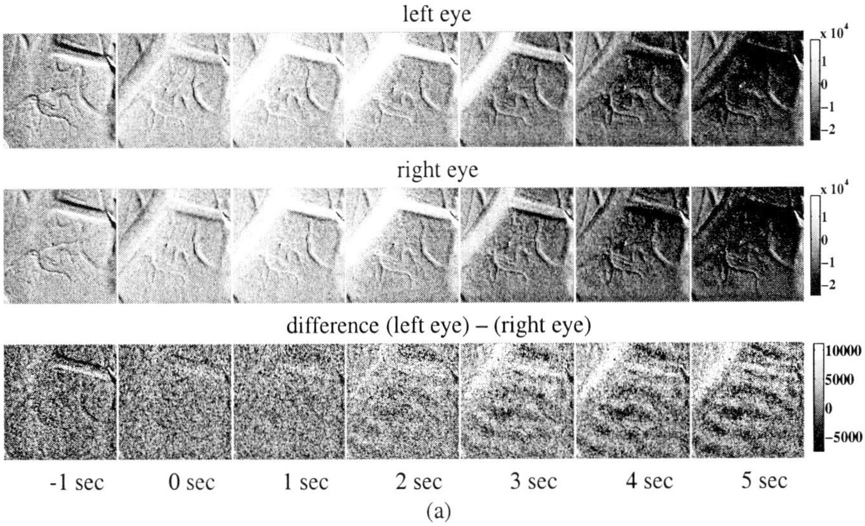


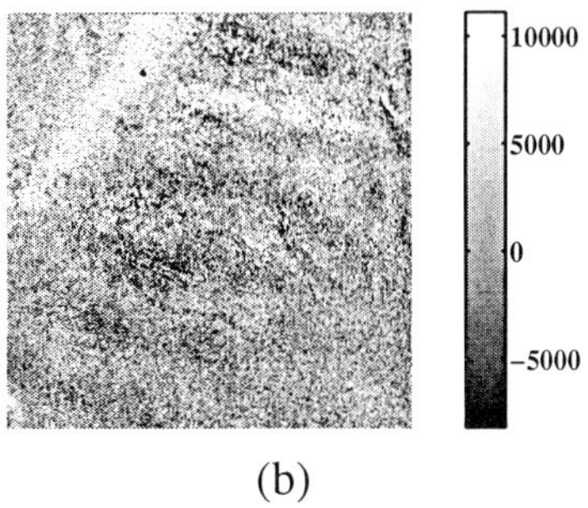


*Figure 5.5.* **(a)** Differential imaging experiment for ocular dominance. Single-condition image stacks for left-eye (top) and right-eye stimulation (center) are subtracted pixelwise from each other to yield the differential image stack (bottom). Blood-vessel patterns and global components have been strongly reduced. **(b)** Differential image obtained by summation of frames 4-7 of the bottom stack in (a).

single-condition stacks can contain signal components which are not locked to the stimulus, but are independent of stimulation. Examples include vasomotion signals (Mayhew et al., 1996a), ongoing activity and detection artifacts. For all these signals, the model assumptions (5.11) and (5.12) are wrong, and differential imaging cannot succeed. Figure 5.6a shows a differential imaging stack after first frame analysis, that has been obtained from the primary visual cortex of a ferret (Stetter et al., 2000c) for 0 deg and 90 deg oriented grating as orthogonal stimuli. Five images (image collection time: 600 ms) were collected during the whole stimulus presentation time of 3 sec (ora 2001 imager, Optical Imaging Inc.) and reduced to a stack of four images by first frame analysis. The stack contains a strong artifact with a roughly annular shape combined with a gradient in gray level from bottom to top, which might be caused by movements in the setup. The artifact is time-dependent, and could not be eliminated by differential imaging (figure 5.6b), because the time courses in the two single-condition stacks are different from each other.

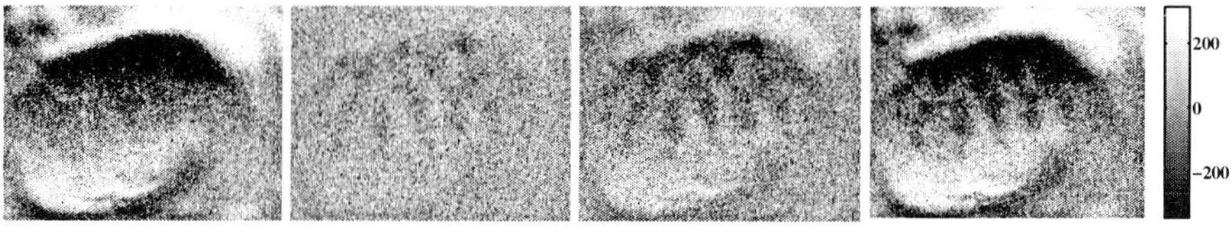


(a)

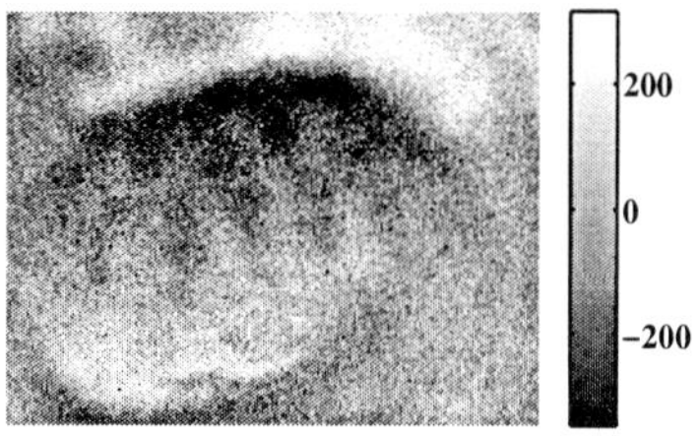


(b)

*Figure 5.6.* **(a)** Differential imaging stack after first frame analysis from ferret area 17 for 0 and 90 deg oriented grating stimuli. **(b)** 0 - 90 deg differential image obtained as the mean of (a). A time-dependent artifact is superimposed on the mapping signal (vertical elongated blob patterns) which cannot be removed by differential imaging. Image specifications: 192 × 144 pixels, 36 $\mu$m per pixel; image area: 6.9 × 5.2 mm) .

A much more serious drawback of differential imaging results from the fact, that we cannot expect to get orthogonal stimulus pairs for every aspect of cortical function we want to explore. Generally, in order to be able to design orthogonal pairs of stimuli, we have to know in advance, which feature selectivities are shown by single neurons. Orientation selectivity of single cells has led to orthogonally oriented gratings, and ocular dominance to pairs of monocular stimuli as orthogonal pairs. There are a variety of tasks for optical imaging, however, for which orthogonal pairs of stimuli are difficult to construct or even do not exist. For example, the measurement of cortical contrast-response curves, the measurement of responses to localized stimuli or the measurement of contextual effects represent experiments, for which orthogonal stimuli do not exist. A design which answers the question, if spatial frequency selectivity in cortex is binary or continuous (Hübener et al., 1997), is difficult to address by differential imaging, but here the reason is that we lack the prior knowledge of what is the nature of spatial frequency selectivity (binary-like or continuous) of cortical neurons. In summary, if we want to address more sophisticated questions than the measurement of (well-known) orientation and ocular dominance maps by optical imaging, we must be able to extract the mapping signal from single condition stacks such as figure 5.2a, top, rather than from differential imaging stacks. Motivated by these considerations, we will score further methods for analyzing optical imaging records also for their ability to extract the mapping signal from single-condition image stacks.

## 2.2 Cocktail-Blank for Single-Condition Imaging

One frequently used procedure for the processing of single-condition images is based on the "cocktail blank". This method can be applied, if we have access to images for a set $\mu = 1, ..., p$ of different stimulus conditions, and if we assume that the sum of the cortical activity patterns for all stimulus conditions covers the cortex uniformly. If this assumption holds and if optical signals superimpose linearly, we can obtain an image which corresponds to the optical recording from a completely and uniformly excited cortex by summing up the responses to all stimulus conditions:

$$x_c(\mathbf{r}) = \frac{1}{p} \sum_{\mu=1}^{p} \sum_{m=1}^{M} x_m^{(\mu)}(\mathbf{r}). \tag{5.17}$$

This summed frame is defined as the cocktail blank. If we additionally assume again, that the global signal components $s_l, l = 2, ..., L$, are identical for all stimulus conditions, the cocktail blank can be written as

$$x_c(\mathbf{r}) = \frac{1}{Mp} \sum_{\nu=1}^{p} \sum_{m} \left( \sum_{l=2}^{L} a_{m,l}\, s_l(\mathbf{r}) + a_{m,1}\, s_1^{(\nu)}(\mathbf{r}) + n_m^{(\nu)}(\mathbf{r}) \right), \tag{5.18}$$

$$= \sum_{l=2}^{L} \bar{a}_l\, s_l(\mathbf{r}) + \bar{a}_1\, \frac{1}{p} \sum_{\nu=1}^{p} s_1^{(\nu)}(\mathbf{r}) + \frac{1}{Mp} \sum_{m} \sum_{\nu} n_m^{(\nu)}(\mathbf{r}) \tag{5.19}$$

where the mapping signal has been assumed again as the first source in the sum. If we subtract the cocktail blank from the $\mu$-th raw single-condition image, we arrive at

$$\hat{s}_1^{\mu}(\mathbf{r}) = \frac{1}{M} \sum_{m} x^{(\mu)}(\mathbf{r}) - x_c(\mathbf{r}) \tag{5.20}$$

$$= \bar{a}_1 s_1^{(\mu)}(\mathbf{r}) - \bar{a}_1 \frac{1}{p} \sum_{\nu=1}^{p} s_1^{(\nu)}(\mathbf{r}) + \frac{1}{Mp} \sum_{m} \sum_{\nu \neq \mu}^{p} n_m^{(\nu)}(\mathbf{r}) \tag{5.21}$$

If all single-condition responses together sum up to a uniform cortical excitation, the second term in eq. (5.21) is constant over space, $\frac{1}{p} \sum_{\nu=1}^{p} s_1^{(\nu)}(\mathbf{r}) = s_c$, and after correction by the cocktail blank, $\hat{s}_1^{(\mu)}(\mathbf{r})$ represents an unbiased estimate of the true single condition response $s_1^{(\mu)}(\mathbf{r})$ up to a constant.

### 2.2.1 Limitations

In general we have to face the fact that cocktail blank images do not lead to true single-condition mapping signals, but always contain contributions from different single condition responses also. This becomes more obvious, if we reformulate eq. (5.21) (using the abbreviation $n'(\mathbf{r})$ for the noise term) as

$$\hat{s}_1^{(\mu)}(\mathbf{r}) = \bar{a}_1 \left( s_1^{(\mu)}(\mathbf{r}) \left(1 - \frac{1}{p}\right) - \frac{1}{p} \sum_{\nu \neq \mu,}^{p} s_1^{(\nu)}(\mathbf{r}) \right) + n'(\mathbf{r}), \qquad (5.22)$$

which for the particular case of a pair of orthogonal stimuli (1) and (2) reduces to the estimator for the differential signal

$$\hat{s}_1^{(\mu)}(\mathbf{r}) = \frac{1}{2}\bar{a}_1 \left( s_1^{(1)}(\mathbf{r}) - s_1^{(2)}(\mathbf{r}) \right). \qquad (5.23)$$

The equations demonstrate that the estimator for a single condition image $\hat{s}_1^{(\mu)}(\mathbf{r})$ always contains contributions from other maps $s_1^{(\nu)}(\mathbf{r}), \nu \neq \mu$. Particularly if the assumptions that lead to the application of the cocktail blank method are not strictly fulfilled, subtraction of or division by the cocktail blank mixes the responses to different stimulus conditions in an uncontrollable way. This method is widely used but its value for obtaining true single condition maps is questionable.

## 2.3 Bandpass Filtering

A further assumption about the different signal components that has been frequently used in the optical imaging literature is bandpass filtering of images (Blasdel, 1992a; Bonhoeffer and Grinvald, 1996). Source separation by bandpass filtering is based on the assumption that the spatial patterns of different sources, $s_l(\mathbf{r})$, are concentrated around different spatial frequencies, i.e. they have different characteristic wavelengths. This assumption can be formulated more quantitatively in the frequency domain as follows: The discrete Fourier transform of a source image $s_l(\mathbf{r})$ is given by

$$S_l(k_x, k_y) = \sum_{r_x=0}^{P_x-1} \sum_{r_y=0}^{P_y-1} s_l(r_x, r_y) \exp\left(-2\pi i \left(\frac{k_x r_x}{P_x} + \frac{k_y r_y}{P_y}\right)\right) \qquad (5.24)$$

and the inverse transform by

$$s_l(r_x, r_y) = \frac{1}{P_x P_y} \sum_{k_x=0}^{P_x-1} \sum_{k_y=0}^{P_y-1} S_l(k_x, k_y) \exp\left(2\pi i \left(\frac{k_x r_x}{P_x} + \frac{k_y r_y}{P_y}\right)\right). \qquad (5.25)$$

We denote the Fourier transform of an image, $s_l(\mathbf{r})$ by the corresponding capital letter. The absolute value of the wave-vector $\mathbf{k}$ specifies the inverse wavelength or spatial frequency of a sine-wave, and its direction is orthogonal to the wave front. For many considerations it is important to know, how strong the contribution of a given frequency component $k = |\mathbf{k}|$ to the source image is irrespective

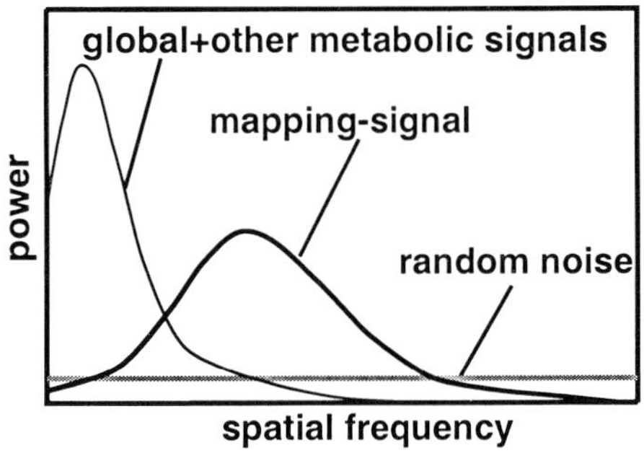


*Figure 5.7* Illustration of typical frequency courses for different signal sources. Even if sources are concentrated around different frequencies, they cannot be generally assumed disjunct.

of its direction of propagation. Using polar coordinates for the wave vector, $\mathbf{k} = (k, \phi)$ with the azimuth angle $\phi$, a measure of this quantity is given by the power spectrum

$$S_l(k) = \int_0^{2\pi} k \, d\phi \; S_l(\mathbf{k}) S_l^*(\mathbf{k}), \tag{5.26}$$

where the superscript * denotes the conjugate complex value of a complex number.

For source separation by bandpass filtering, the model assumption is that different sources are concentrated in disjunct frequency bands of the power spectrum. This assumption was motivated by the following observations: $(i)$ For ocular dominance and orientation selectivity the global signals are concentrated around lower spatial frequencies than the mapping signal (cf. chapter 4, section 1.3 and figure 5.5a). $(ii)$ White noise contains components with high spatial frequency, which are not present in images of intrinsic signals due to metabolic spread and optical blur by tissue scattering (cf. chapter 4, section 3). Figure 5.7 illustrates the expected frequency courses of different signal sources. White noise is flat in frequency space, and dominates the power spectrum at high spatial frequencies, whereas the global signal and other metabolic signals might occupy mainly the low spatial frequency domain. The mapping signal is expected to occupy an intermediate range of frequencies.

Given these model assumptions, we can exclude the low and high spatial frequencies of the record by a bandpass filter, and thereby can reduce the contamination of the mapping signal by the other components. A bandpass filter can be easily applied in the frequency domain by multiplication with the circularly symmetric function

$$B(\mathbf{k}) = f\left(k_{\text{lp}} - |\mathbf{k}|\right) \, f\left(|\mathbf{k}| - k_{\text{hp}}\right), \tag{5.27}$$

$$f(x) = \frac{1}{1 + \exp(-\beta x)}. \tag{5.28}$$

Figure 5.8 demonstrates the application of a bandpass filter to the differential image 5.8a (same as in figure 5.6b). The mapping signal is contaminated by a

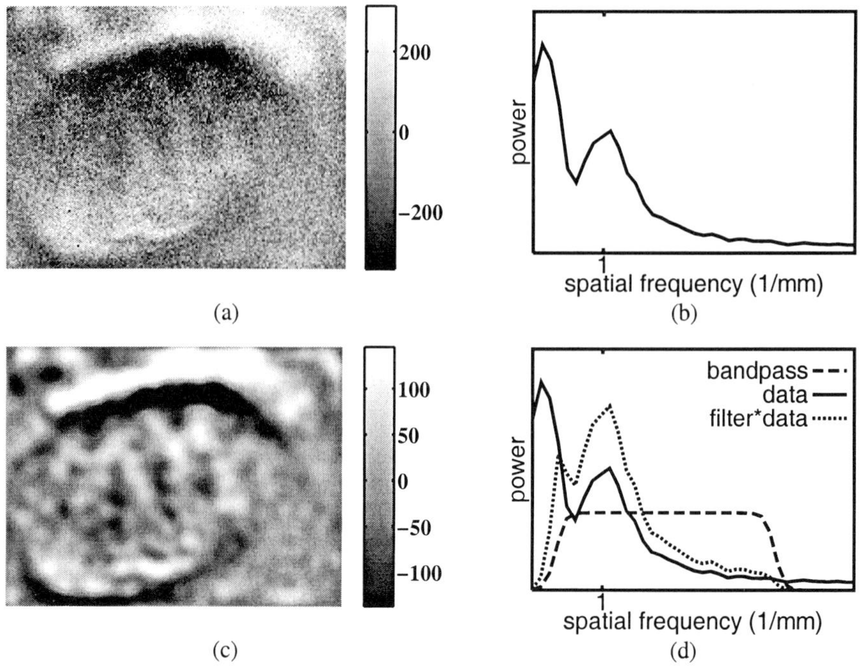


*Figure 5.8.* Source separation by bandpass filtering.**(a)** Original 0-90 deg differential image from ferret area 17 (same as in figure 5.6b). **(b)** Power spectrum of the differential image shown in (a). **(c)** Same differential image after bandpass-filtering (parameters: highpass cutoff: $k_{\rm hp} = 0.4$ mm$^{-1}$; lowpass cutoff: $k_{\rm lp} = 3.26$ mm$^{-1}$; $\beta = 1.4$ mm). **(d)** The original power spectrum (solid line), the profile of the bandpass filter applied (dashed line) and the power spectrum of the differential image after bandpass-filtering (dotted line).

global annular artifact combined with a gradual increase of the gray-level from top to bottom. Figure 5.8b shows the power spectrum of the image, which has been obtained by embedding the zero-mean image into a $256 \times 256$ frame, application of a fast Fourier transform (FFT) algorithm (Press et al., 1988), and subsequent application of eq. (5.26). In this case, two different signal components can be clearly distinguished from the power spectrum: One that is concentrated at low spatial frequencies and another concentrates at intermediate spatial frequencies. Towards the high frequency end the spectrum becomes flat and is dominated by white noise. Figure 5.8c shows the image after application of a bandpass filter, which partially removes the low spatial frequency components and the white noise (the corresponding power spectrum and the applied bandpass filter is shown in figure 5.8d). The central region of the image has been flattened by the reduction of the global component, and the elongated blobs of the mapping signal have been emphasized a little bit with respect to the noise, but a complete and deterioration-free extraction of the mapping signal was not achieved.

### 2.3.1 Limitations

Bandpass filtering also suffers from a number of drawbacks. In summary: $(i)$ Sources are not disjunct in frequency space, and their behavior depends on the experiment. $(ii)$ The filter values are arbitrary and strongly affect important properties of the resulting images. $(iii)$ Bandpass-filtered noise looks similar to many optical maps. Signals and noise are therefore difficult to distinguish from each other after bandpass filtering.

Figure 5.8 already demonstrates, that even the global and the mapping component, which motivated post-processing by bandpass-filtering, are not disjunct in frequency space, but overlap. Even an optimum filter with a highpass cutoff located at the local minimum between both peaks, cannot perform a complete separation. Either the global signal is incompletely removed, or the mapping signal becomes adulterated, or a combination of both effects happens. Additionally, we do not know what the true frequency course of each signal is: Is the low-frequency peak really only due to global components? How far do global components reach towards the high frequency end? Furthermore the spectrum of the mapping signal depends on the stimulus condition and on the aspect of cortical representation we want to investigate. It may be concentrated at intermediate frequencies for orientation selectivity, but concentration is already questionable for stripe-like ocular dominance patterns, and the concentration may be at completely different frequencies depending on the stimulus applied. For example, responses to positional stimuli, which are used for the exploration of retinotopic order, may have typical wavelengths of several millimeters (McLoughlin and Blasdel, 1998). Other components, namely blood-vessel patterns show even stronger overlap with the mapping signal than the global signal. Blood vessels are strongly elongated structures and contain both high- and low-frequency components, and bandpass filtering fails to remove them from the data.

Figure 5.9 demonstrates this drawback for the example of the single-condition stack shown in figure 5.2, from which the single-condition image for left-eye stimulation (figure 5.9a) has been obtained by pixelwise summation. The horizontal stripe pattern formed by the mapping signal is barely visible in the presence of the strong blood vessel patterns as expected. In figure 5.9b, the same image after bandpass filtering is displayed. Low spatial frequency components including the bright region within the large blood vessel in the top part of the image and the global darkening below that blood vessel have been reduced, and the image has been smoothened. However, the blood-vessel patterns are still present in the image, and still are much stronger than the mapping signal.

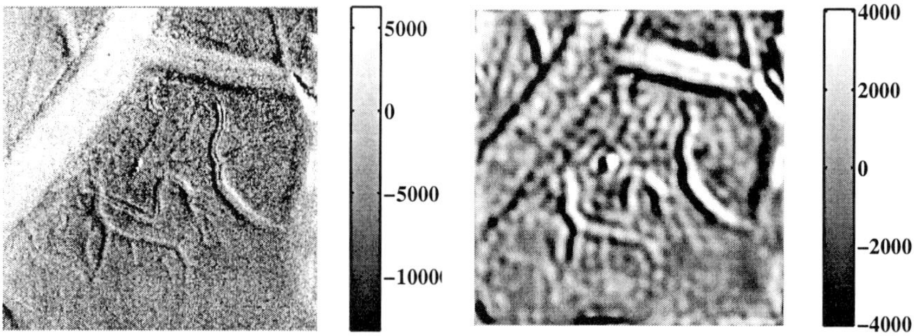


(a) (b)

*Figure 5.9.* **(a)** Single-condition image obtained as the pixelwise sum over the single-condition stack shown in figure 5.2. **(b)** Same image after bandpass filtering. Though global reflectance changes (bright blood vessel, global darkening) are reduced, the blood-vessel patterns still dominate the picture. Highpass cutoff: 1.06 $mm^{-1}$ ($\lambda = 0.95$ mm); Lowpass cutoff: 6.6 $mm^{-1}$ ($\lambda = 0.15$ mm);

The application of bandpass filtering raises the question, which filter values shall be used for filtering. If the power spectrum of an image is multimodal such as in figure 5.8b, we can take the highpass cutoff frequency, which corresponds to the local minimum between the peaks. If the total spectrum consists of a linear superposition of several mono-modal spectra for the individual sources, and if we define the separation error as the summed fractions of the spectra that are mistreated (e.g., the fraction of the mapping signal that is cut away plus the fraction of the global signal that is preserved), the choice of the cutoff frequency at the local minimum will minimize the separation error (Bishop, 1995). However, many images do not contain expressed minima, and the cutoff frequencies of the filters – within some reasonable intervals – have to be treated as arbitrary parameters. This observation raises the question, if and how observables of the functional cortical organization, which we want to extract from optical imaging measures, do depend on the choice of the cutoff frequencies.

Figure 5.10 demonstrates, that the quality of separation as well as important properties of cortical activity maps depend critically on the arbitrary choice of the high-pass cutoff frequency. The figure displays the number of singularities (solid line) of the orientation map (cf. figure 3.16a) optically recorded from the visual cortex of a ferret (43d old) as a function of the cutoff-frequency of the high-pass filter. The insets are orientation histograms obtained after analysis without (left) and with (right) highpass filtering. They show, how frequently pixels of a given preferred orientation (in bins of 15 deg) are found within the region of interest of the orientation map. The dash-dotted and dashed lines plot the first and the second Fourier components, respectively, of the

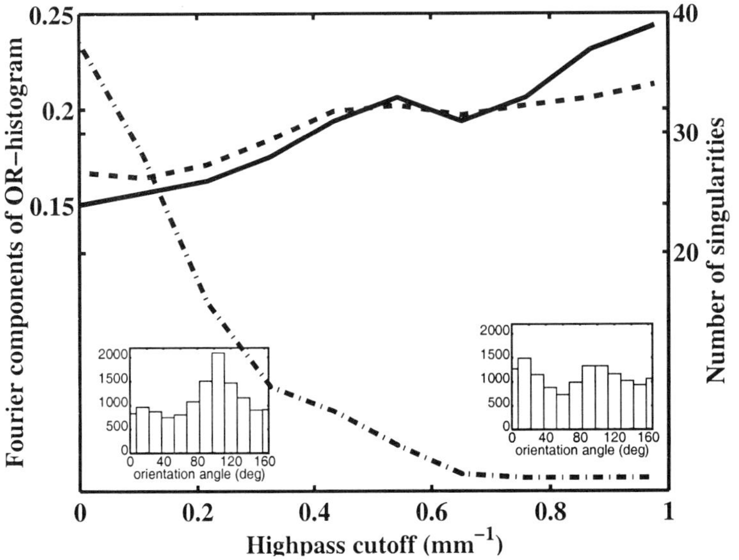


*Figure 5.10.* Biologically relevant quantities which are derived from the orientation map of ferret visual cortex as a function of the cutoff frequency of an applied high-pass filter. The diagram plots the number of singularities (solid line) and the first (dash-dotted line) and second (dashed line) Fourier components of the orientation histogram as a function of the cutoff frequency. The low-pass cutoff was 2.7 mm$^{-1}$ ($\lambda$ = 0.37 mm). Insets show the orientation histograms without high-pass-filtering (left) and after high-pass-filtering (right) with the cutoff-frequency 0.54 mm$^{-1}$ ($\lambda$ = 1.84 mm). All quantities depend on the choice of the high-pass cutoff frequency (behavior shown is typical).

orientation histogram as functions of the cutoff-frequency of the high-pass filter. These numbers represent measures for the shape of the orientation histogram: A monomodal and peaked orientation histogram is indicated by a large first Fourier component, whereas a bimodal histogram would give rise to a large second Fourier component. All quantities depend on the choice of the cutoff-frequency for which there is no way to determine a "correct" value. Observables turn out to be less sensitive against the choice of the lowpass-cutoff frequency as long as the filter does not remove considerable parts of the mapping signal.

Bandpass filtering of optical maps introduces a further difficulty: Particularly if aspects of local cortical processing are investigated by optical imaging, post-processing by bandpass filtering introduces a possible source of ambiguity in the data. Many aspects of cortical organization, which we want to detect, are arranged in periodic patterns. But the same is true for bandpass-filtered noise: its periodicity is determined by the frequency band that is preserved by the filter. This observation has even triggered structural models, which use bandpass-filtered noise for a phenomenological description of the structure of

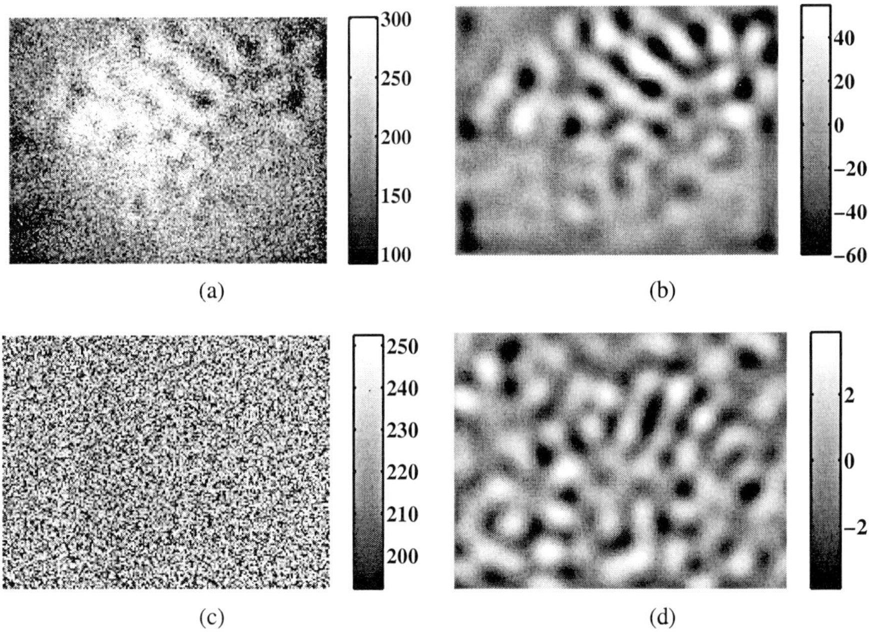


*Figure 5.11.* **(a)** 0-90 deg differential image from ferret primary visual cortex (same record as used for figure 5.10), and **(b)** the same image after bandpass filtering (highpass cutoff: 0.43 $\mathrm{mm}^{-1}$ ($\lambda = 2.3$ mm); Lowpass cutoff: 1.63 $\mathrm{mm}^{-1}$ ($\lambda = 0.61$ mm)). **(c)** Random white noise image with same mean and variance than **(a)**, and **(d)** its bandpass-filtered version (same cutoff frequencies). Though the signal is weaker in the bandpass filtered image, its spatial structure is very similar to the differential image.

orientation maps in area 17 (Rojer and Schwartz, 1990). Thus, bandpass-filtered noise can be very similar to the signals we want to measure. As a consequence, signal patterns and noise patterns may become very similar to each other in a bandpass filtered image, and it may be difficult to distinguish them from each other. Figure 5.11 demonstrates this effect for the same data set as used in figure 5.10. The top row shows an original 0-90 deg differential image (left) and its bandpass-filtered version, whereas the bottom row shows a random white noise pattern with the same mean and variance than the differential image before (left) and after bandpass-filtering with the same filter values. A comparison between both bandpass-filtered versions demonstrates the expressed similarity between the noise and the differential imaging patterns. In this example, the noise variance after bandpass filtering is much lower than the variance of the optical signal (see scale bar), but generally the dB-signal-to-noise ratio can be close to zero in optical imaging experiments, and in this case even the signal sizes would become comparable to each other.

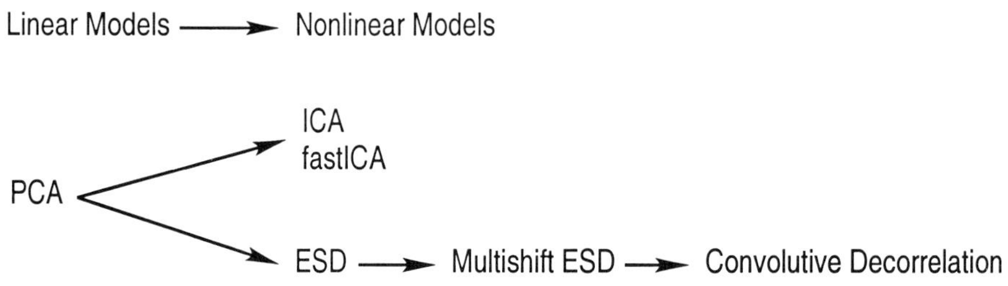


*Figure 5.12.* Overview over source separation methods which we have applied to optical imaging data together with their mutual relationships. The methods are based on different assumptions or sources of knowledge about the signals including knowledge about functional relationships (top) and assumptions about statistical properties of sources (bottom). PCA: principal component analysis. ICA: independent component analysis. ESD: extended spatial decorrelation.

Part of the problem described here can be overcome shifting the center frequency of the bandpass filter away from the dominant frequency of the mapping signal. The effect of this procedure can be seen in figure 5.8, where the filtered noise (outside the annular artifact) appears more wiggly than the mapping signal inside.

The present chapter can be summarized as follows: We have formulated the data analysis task for optical imaging as a source separation problem and have described three heuristic methods, which have been established in the literature as standard tools for the solution of this problem. We have demonstrated, that these heuristic methods can be viewed within the framework of the source separation problem as formulated in section 1.2, and that they are based on the following particular assumptions about the behavior of different signal components: Differential imaging assumes completely stimulus-unspecific global components and an ideally stimulus-specific mapping signal, whereas normalization by a cocktail blank in addition assumes that a set of stimulus-condition as a whole excites the cortex uniformly. Bandpass-filtering, in contrast, is based on the assumption that different signal components cover different regions in the frequency domain.

We and many others (Blasdel and Salama, 1986; Grinvald et al., 1986; Ts'o et al., 1990; Bonhoeffer and Grinvald, 1991; Blasdel, 1992a; Kisvarday et al., 1994; Bonhoeffer and Grinvald, 1996; Sengpiel et al., 1998b) have demonstrated that these methods can work well for selected real data sets. At the same time it has become clear that for many optical imaging experiments (actually the majority of all sets), these assumptions may not be true and the application of differential imaging or bandpass filtering may fail or even yield misleading results. It is therefore reasonable and necessary to search for alternative approaches for the analysis of functional brain imaging data. These methods should be applicable to the source separation problem eq. (5.7) or (5.10) and should be able to take reasonable assumptions about the different sources as input. Two classes of methods, which are summarized in figure 5.12, have been successfully applied to the analysis of optical imaging data: Regression methods (top) (Mayhew et al., 1996b; Stetter et al., 1997c; Greve et al., 1999a; Greve et al., 1999b; Stetter et al., 2000b), which describe data sets as combinations of deterministic model functions and noise, and projection methods (bottom) (Stetter et al., 1997c; Everson et al., 1997; Otto et al., 1998; Schießl et al., 1999; Schöner et al., 2000; Schießl et al., 2000b; Stetter et al., 2000c; Vollgraf et al., 2000; Schießl et al., 2000a), which try to find "interesting" directions in the data space according to a given criterion. In the following two chapters we will describe regression methods and projection methods, respectively, and will demonstrate on artificial data sets how they can be applied to image stacks. Chapter 8 will provide selected applications of the statistical methods to real data sets.

Chapter 6

# REGRESSION METHODS FOR SOURCE SEPARATION

In many statistical analysis tasks, we are confronted with data sets, which are arranged in pairs of data points $(t_m, x_m)$, $m = 1, ..., M$. For example, $x_m$ may be some observable which has been measured at times $t_m$, and we wish to characterize the statistical structure in this data set. Often, there is some reason to believe that there is an underlying deterministic functional relationship between the pairs of data. The characterization of this underlying functional relationship based on the data set $\{(t_m, x_m)\} \equiv (t_m, x_m)$, $m = 1, ..., M$, represents a regression task (Kay, 1993; Bishop, 1995).

Formally, we assume that each dependent data point, say $x_m$, is connected to its partner $t_m$ via a function $a(t; \mathbf{w})$, i.e. it represents a noisy sample of the deterministic function $a(t_m; \mathbf{w})$ at $t_m$:

$$x_m = a(t_m; \mathbf{w}) + n_m, \quad m = 1, ..., M. \tag{6.1}$$

In eq. (6.1), $n_m$ are realizations of some random noise variable. $a(t; \mathbf{w})$ denote a class of functions which are parameterized by the vector of statistical parameters $\mathbf{w} = (w_1, ..., w_L)^T$, which is unknown and must be estimated from the data. Because the parameter vector exhaustively describes all properties of the model function, we will often refer to $\mathbf{w}$ itself as the model. In the following sections, we will consider special cases of this regression scheme and will show how they can be applied for source separation in imaging data.

## 1. General Linear Models

A general linear model (GLM) represents a regression method, for which a special assumption about the function $a(t, \mathbf{w})$ is made: the function is assumed

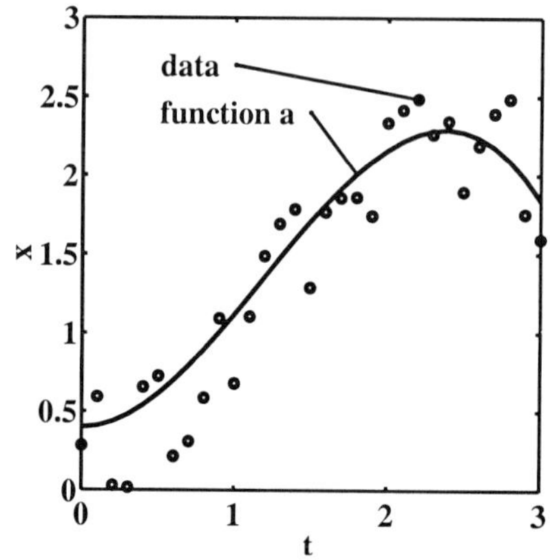


*Figure 6.1* Illustration of a regression task: characterization of the underlying functional relationship from a set of data pairs.

to depend linearly on the model parameters $\mathbf{w}$:

$$a(t;\mathbf{w}) \equiv \sum_{l=1}^{L} a_l(t) w_l = \mathbf{a}(t)^T \mathbf{w}. \tag{6.2}$$

This means that a general linear model describes the output data as a linear combination of fixed functions $a_l(t),\ l = 1, ...L$ plus additive noise:

$$x_m = \sum_{l=1}^{L} a_l(t_m) w_l + n_m, \quad m = 1, ..., M \tag{6.3}$$

Note that this does by no means imply that the model functions themselves are linear, but instead that they are linearly mixed to explain the data. However eq. (6.3) implies that the functions $a_l(t)$ are assumed to be known, and the $L$ parameters $w_l$ encode the amplitudes of the functions. Estimating the parameters means to estimate, how much each of the model functions contributes to the observed relationship between $t$ and $x$. Figure 6.2 illustrates, how a general linear model works.

The model eq. (6.2) can be written in matrix form as

$$\mathbf{x} = \mathbf{A}\mathbf{w} + \mathbf{n}, \tag{6.4}$$

where the data points and noise have been combined to vectors $\mathbf{x} = (x_1, ..., x_M)^T$ and $\mathbf{n} = (n_1, ..., n_M)^T$, respectively. The $M \times L$ matrix $\mathbf{A}$ contains the samples of the model functions as columns,

$$\mathbf{A} = \begin{pmatrix} a_1(t_1) & a_2(t_1) & \cdots & a_L(t_1) \\ a_1(t_2) & a_2(t_2) & \cdots & a_L(t_2) \\ \vdots & \vdots & & \vdots \\ a_1(t_M) & a_2(t_M) & \cdots & a_L(t_M) \end{pmatrix} \tag{6.5}$$

and is called the *design matrix* of the GLM.

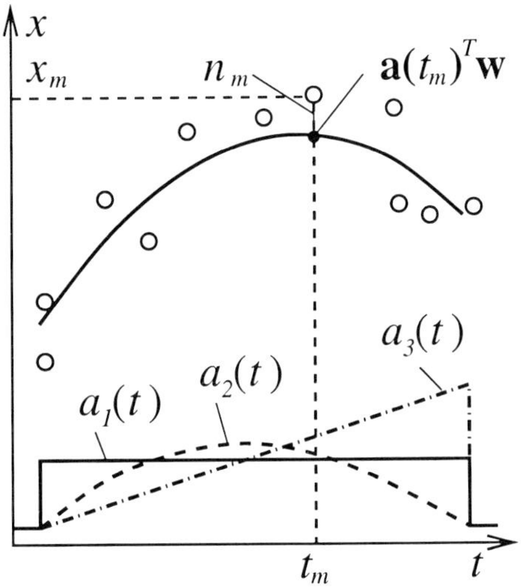


*Figure 6.2* Principle of a general linear model: A set of data pairs $(t_m, x_m)$ is described as a linear combination of model functions $a_l(t)$ plus a random noise variable $n_m$ (the example shows three model functions). The linear model parameters $\mathbf{w}$ describe the amplitudes of the functions, their curve shapes are assumed to be known.

## 1.1 Maximum Likelihood Estimation of Parameters

We wish to find an optimal set of parameters, for which the linear model eq. (6.3) describes the data best. For this we have to define what is a good description of the data. One common measure for the quality of a regression is based on the criterion, how likely the observed set of output data, $\mathbf{x}$, has been generated by a given model $\mathbf{w}$ on the basis of the input data $\mathbf{t} = (t_1, ..., t_M)$. This measure can be written as the conditional probability

$$p(x_1, ..., x_M \mid t_1, ..., t_M; \mathbf{w}) \tag{6.6}$$

and is called the *likelihood* of the data given the model. If we assume that different pairs of data are drawn independently from the underlying distribution, the likelihood factorizes and can be written as

$$p(x_1, ..., x_M \mid t_1, ..., t_M; \mathbf{w}) = \prod_{m=1}^{M} p(x_m \mid t_m; \mathbf{w}) \tag{6.7}$$

An expression for the likelihood can be obtained using the following consideration: Given the model assumption eq. (6.3), we can identify any contribution to the data, which cannot be explained by the model, as noise:

$$n_m = x_m - \sum_{l=1}^{L} a_l(t_m) w_l. \tag{6.8}$$

The probability for a data pair $(t_m, x_m)$ to be generated by a given model $\mathbf{w}$ is the higher, the more likely it is, that the mismatch between model and data

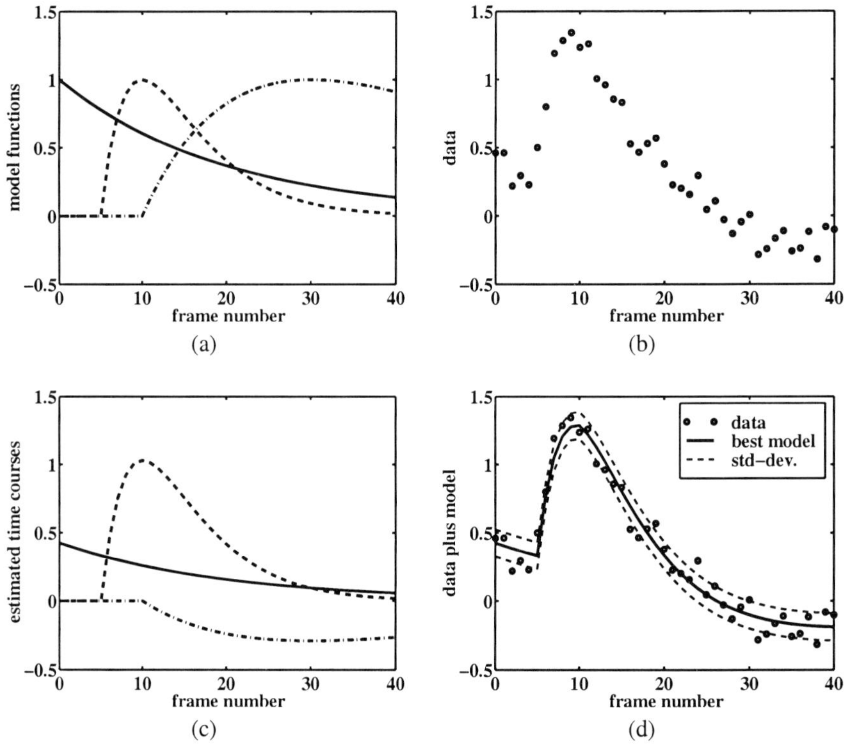


*Figure 6.5.* Demonstration of a fit using a GLM. **(a)** Three model functions **(b)** Data set as noisy mixture of the functions in (a). **(c)** Best fits of the signal time courses using a GLM with the true model functions (a). **(d)** Data (circles) plus best fit (solid line). The dashed lines mark the distance of one estimated noise standard deviation $\sigma_n$ to either side.

Finally, because application of the GLM is a local operation in space, we can abbreviate eqs. (6.27) and (6.28) by using the data matrix notation eq. (5.7) to

$$\mathbf{X} = \mathbf{A}\mathbf{S} + \mathbf{N}, \tag{6.30}$$

$$\hat{\mathbf{S}} = (\mathbf{A}^T\mathbf{A})^{-1}\mathbf{A}^T\mathbf{X} \tag{6.31}$$

where $\mathbf{S}$ is the $L \times P$ matrix which contains the $L$ source images as rows.

## 1.4 Examples

Figure 6.5 shows a simple example based on an artificial pixel time series, which demonstrates how a GLM works. In figure 6.5a, three model functions are plotted – two alpha-functions $x(t) = ((t-\tau)/\delta)\exp(-(t-\tau)/\delta)$ with $(\tau_1, \delta_1) = (10, 20)$ (dash-dotted line) and $(\tau_2, \delta_2) = (5, 5)$ (dashed line) and one exponential function $x(t) = \exp(-t/\tau_3)$ with $\tau_3 = 20$ (solid line) –

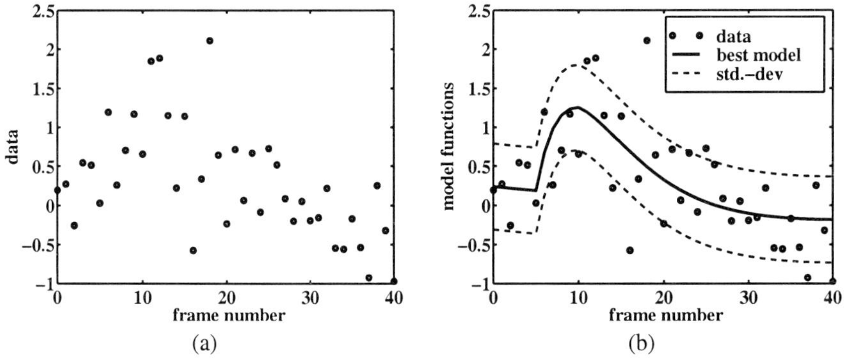


*Figure 6.6.* Fit of strongly noisy data by a GLM **(a)**. Data from same model functions and mixing coefficients as in figure 6.5, but with stronger noise, $\sigma_n = 0.5$. **(b)** Best fit using a GLM and the true functions.

which have been used and combined with Gaussian white noise with a standard deviation of $\sigma_n = 0.1$ to generate the noisy time series in figure 6.5b. Figure 6.5c shows the same data set (circles) together with the best fit using a GLM, which has been obtained as $\hat{\mathbf{x}} = \mathbf{A}\hat{\mathbf{s}}$ (solid line). The dashed lines mark the estimated standard deviation of the noise $\hat{\mathbf{x}} \pm \hat{\sigma}_n$. In the example shown, the true amplitudes of the three curves were $\mathbf{s} = (-0.3, 1, 0.5)$, and their estimates were obtained by the GLM as $\hat{\mathbf{s}} = (-0.2949 \pm 0.0244, 0.999 \pm 0.046, 0.491 \pm 0.045)$, where the errors denote the standard deviations of the amplitude estimates according to eq. (6.25). The estimated noise amplitude was $\hat{\sigma}_n = 0.101$. All estimated parameters are very close to the true values, and the time series is well described by the model.

Given the assumptions are true, a GLM works well even for very low signal-to noise levels. Figure 6.6 demonstrates the noise robustness of a GLM for the same set of model functions and mixing coefficients as in figure 6.5, but now the noise level was $\sigma_n = 0.5$. Though the structure of the data has become almost invisible due to the strong noise (figure 6.6a), the model can still reliably recover the amplitudes. Figure 6.6b again shows the best fit (solid line) and the confidence intervals (dashed lines) together with the data. In this case, the estimated amplitudes were $\hat{\mathbf{s}} = (-0.40 \pm 0.16, 0.83 \pm 0.30, 0.81 \pm 0.289)$ resulting in Z-scores near 3, and the estimated noise standard deviation was $\hat{\sigma}_n = 0.64$. The reason why a GLM seems to work so astonishingly well is that we use an extremely strong source of knowledge for the fit, namely the exact shape of the model functions. If we do not use the correct functions for the fit, a GLM may fail completely (cf figure 6.9).

Figures 6.7 and 6.8 demonstrate, how a GLM can be used to extract individual source patterns from an image stack, if the time courses of the patterns are

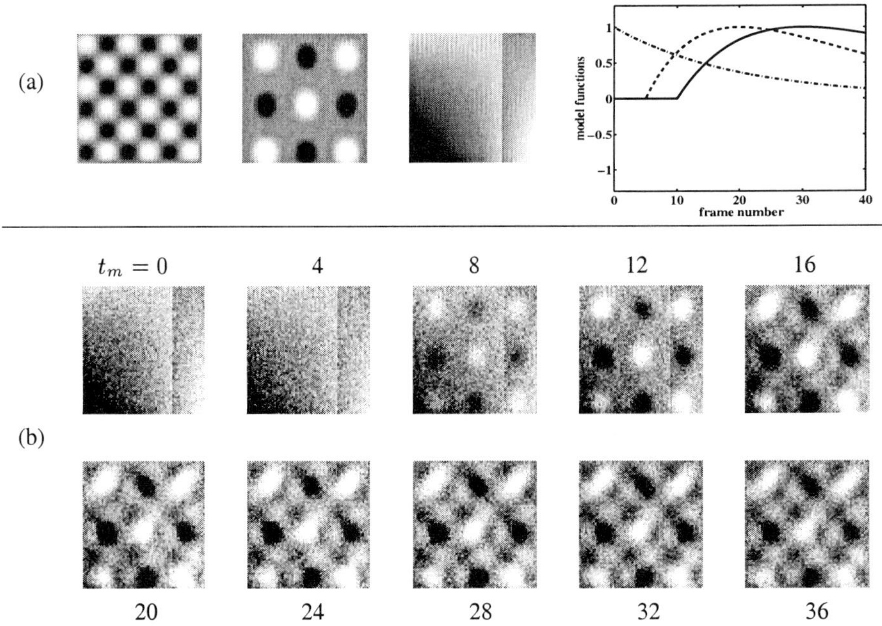


*Figure 6.7.* Artificial image stack as a toy-source separation problem. **(a)** Left: Set of artificial source images $s_l(\mathbf{r})$; Right: their corresponding time courses $a_l(t)$. Solid, dashed and dash-dotted lines correspond to the left, middle and right image, respectively. **(b)** Image stack obtained from the patterns in (a) with amplitudes -0.5,1,1 and $\sigma_n = 0.3$. Every fourth image is shown. Gray scale: $\pm 2$

known. The left part of figure 6.7a displays a set of three artificial source patterns $s_l(\mathbf{r}), l = 1, 2, 3$ with $64 \times 64$ pixels, each normalized to unit variance. The plot at the right part displays three time series $a_l(t), l = 1, 2, 3$, where the first and second functions are alpha functions with $(\tau_1, \delta_1) = (10, 20)$ ($l = 1$, solid line) and $(\tau_2, \delta_2) = (5, 15)$ ($l = 2$, dashed line) and the third function is an exponential decay with $\tau_3 = 20$ sec ($l = 3$, dash-dotted line ). From these images and time series, the image stack shown in figure 6.7b was generated as

$$x(\mathbf{r}, t_m) = -0.5\ a_1(t_m)s_1(\mathbf{r}) + a_2(t_m)s_2(\mathbf{r}) + a_3(t_m)s_3(\mathbf{r}) + n_m(\mathbf{r}) \quad (6.32)$$

with $t_m = 0, 1, ..., 40$ and a noise standard deviation of $\sigma_n = 0.3$. There are frames which almost purely contain the second or third source, but in particular the first source is always strongly mixed with the two other patterns source and is barely visible in the image stack. We regard the first source $s_1$ as a model for the mapping signal, which we want to recover from the image stack.

Figure 6.8a shows the statistical parametric maps $\hat{s}_l(\mathbf{r}), l = 1, 2, 3$ that are obtained from the image figure 6.7b stack by application of a GLM eq. (6.28) with the correct model functions. The tree maps reflect the individual source

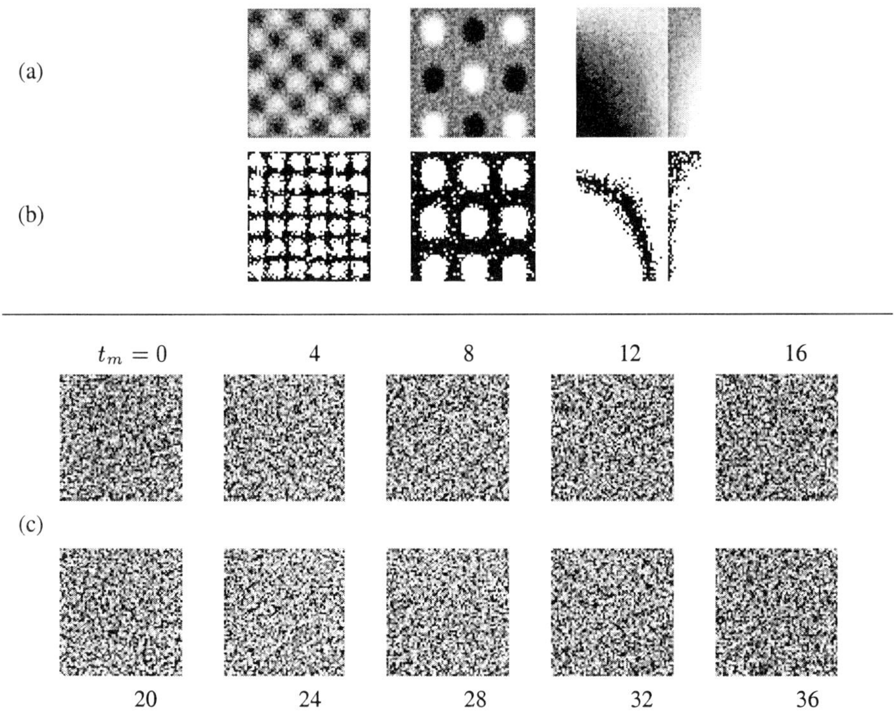


*Figure 6.8.* Source separation using a GLM. **(a)** Statistical parametric maps $\hat{s}_l(\mathbf{r})$ (Gray scale:$\pm 1.5$) obtained from the image stack figure 6.7 using the true model functions. **(b)** Digital Z-score. The regions, where the absolute value of the Z-score exceeds the value of 2 are marked white. **(c)** The residual of the fit (every fourth image shown; gray scale $\pm 0.5$ )

patterns and their amplitudes very well, i.e. all three sources have been very well separated from the noisy mixture by the GLM. The image row 6.8b below the estimated sources provides the digital Z-score for the statistical parametric maps. Every pixel, for which the absolute Z-score eq. (6.26) of the corresponding parameter exceeds 2 shows a bright value. The digital Z-score can be used to mark regions where the corresponding parameter is significantly present. The residual for the fit can be seen in figure 6.8c (every fourth image shown). The residual does not contain any visible structure, which means that the model has extracted all the structure in the data.

This example demonstrates that a GLM is able to solve a linear source separation problem very well, if the model functions are known exactly. This is usually not the case and figure 6.9 demonstrates what happens, if the wrong model functions are used for the fit. This time, the tree model functions shown in figure 6.9a were used for the GLM instead of the true functions 6.7a. The resulting SPMs are shown in the top row of figure 6.9b, the corresponding dig-

(NPM). A nonlinear parametric model helps both restricting the number of parameters and resolving possible ambiguities in the sets of model functions.

## 2.1 Model Framework and Parameter Estimation

A nonlinear parametric model represents a special case of the general regression task eq. (6.1), in which the regression function can be written as a linear combination of parameterized model functions

$$a(t;\mathbf{w}) = \sum_{l=1}^{L} a_l(t;\Phi)s_l, \quad \mathbf{w} \equiv (\Phi, \mathbf{s}). \tag{6.33}$$

In eq. (6.33), $a_l(t;\Phi)$ denote the $L$ families of candidate model functions, the curve shapes of which are parameterized by a set of nonlinear parameters $\Phi = (\phi_1, ..., \phi_c)^T$. This vector can be considered as a collection of parameter subsets, $(\phi_1^1, ..., \phi_{c1}^1, ...\phi_1^L, ..., \phi_{cL}^L)^T$, where the subset $(\phi_1^l, ..., \phi_{cl}^l)^T$ parameterizes the $l$-th family of model functions, i.e. $a_l(t;\Phi) \equiv a_l(t;(\phi_1^l, ..., \phi_{cl}^l))$. The values of $\Phi$ have to be estimated from the data. Because there is a nonlinear relationship between $\Phi$ and the data, these parameters are referred to as nonlinear parameters and the model approach is referred to as a nonlinear model. The vector $\mathbf{s} = (s_1, ..., s_L)^T$ contains the amplitude parameters of the model functions and its components are referred to as linear parameters. They provide the amplitude or signal strength of every component within the time series and must also be estimated from the data. The total parameter set is given by $\mathbf{w} = (\Phi, \mathbf{s})$.

Using a NPM, a given time series of signals, $\mathbf{x} = (x_1, ..., x_M)^T$, can be written as

$$x_m = \sum_{l=1}^{L} a_{m,l}(\Phi)s_l + n_m, \quad m = 1, ..., M, \tag{6.34}$$

where $a_{m,l}(\Phi) = a_l(t_m, \Phi)$. Similarly to eq. (6.4), we can write the model eq. (6.34) in matrix form as

$$\mathbf{x} = \mathbf{A}(\Phi)\mathbf{s} + \mathbf{n}, \tag{6.35}$$

but now the $M \times L$ design matrix $\mathbf{A}(\Phi)$ depends on the nonlinear parameters of the model.

Both linear an nonlinear parameters can be estimated using the principle of maximum likelihood (cf. section 1.2). For Gaussian noise we have to maximize

$$p(\mathbf{x} \mid \mathbf{t};\mathbf{w}) \equiv p(\mathbf{x} \mid \mathbf{t};\Phi,\mathbf{s}) \tag{6.36}$$

$$= \frac{1}{(2\pi\sigma_n^2)^{M/2}} \exp\left(-\frac{\mathbf{n}^T\mathbf{n}}{2\sigma_n^2}\right) \tag{6.37}$$

$$= \frac{1}{(2\pi\sigma_n^2)^{M/2}} \exp\left(-\frac{1}{2\sigma_n^2}(\mathbf{x} - \mathbf{A}(\Phi)\mathbf{s})^T(\mathbf{x} - \mathbf{A}(\Phi)\mathbf{s})\right) \tag{6.38}$$

$$\stackrel{!}{=} \max.$$

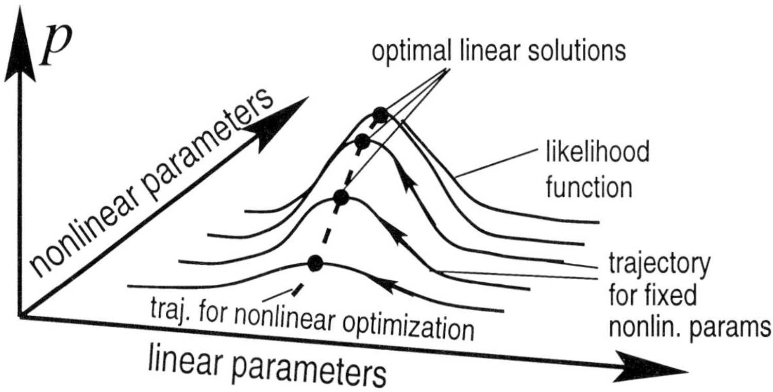


*Figure 6.10.* Simplified maximization of the likelihood function: For each arbitrary but fixed nonlinear parameter, the linear parameters are optimized analytically (black dots). The nonlinear parameters are optimized along the linear optima (dashed line).

Equation (6.38) represents a multidimensional optimization problem, the solution of which is time-consuming. Figure 6.10 illustrates, how we can simplify the maximization by treating linear and nonlinear parameters separately. For this we make use of the fact that the likelihood function is Gaussian in the direction of the linear parameters. For an arbitrary but fixed nonlinear parameter set $\Phi$, the optimization task then reduces to finding the optimal linear parameter set, which can be done analytically using eq. (6.18). In figure 6.10 this means that for every value of the nonlinear parameters, we maximize the likelihood along the direction of the linear parameters, i.e. we find the maximum of the likelihood along the solid lines (marked as a black dot). The nonlinear parameters are then optimized along the trajectory of all optimal linear solutions (dashed line in figure 6.10).

This two-step procedure can be mathematically formulated as follows: For a fixed but arbitrary set of nonlinear parameters, $\Phi$, the optimal linear parameters $\mathbf{s}_{\text{Opt}}(\Phi)$ are determined as:

$$\mathbf{s}_{\text{Opt}}(\Phi) = \left(\mathbf{A}^T(\Phi)\mathbf{A}(\Phi)\right)^{-1} \mathbf{A}^T(\Phi)\, \mathbf{x}. \tag{6.39}$$

The residual is then given by

$$\begin{aligned} \mathbf{n}(\Phi) &= \mathbf{x} - \mathbf{A}(\Phi)\, \mathbf{s}_{\text{Opt}}(\Phi) \\ &= \left(\mathbf{I} - \mathbf{A}(\Phi)\left(\mathbf{A}^T(\Phi)\mathbf{A}(\Phi)\right)^{-1} \mathbf{A}^T(\Phi)\right) \mathbf{x}. \end{aligned} \tag{6.40}$$

values $\Phi_0 = (12, 5, 3, 3, 20)$ for the NPM. Because the model functions are not correct, a GLM based on these functions will yield a suboptimum fit, which is plotted as a solid line in figure 6.11b. Figure 6.11c provides the best set of model functions after optimization of the curve shapes by a nonlinear model (Optimization was carried out for the parameters $\tau_1, \delta_1, \tau_2$ and $\delta_2$). The cost function was minimized using ordinary gradient descent with adaptive step size control (Stetter et al., 2000b) (131 iterations, stopping criterion: less than $10^{-5}$ change in the residual) and yielded $\Phi_{\text{Opt}} = (11.45, 13.17, 4.97, 5.75, 20)$ and $\hat{\mathbf{s}} = (-0.38 \pm 0.11, 0.92 \pm 0.17, 0.54 \pm 0.14)$, which is reasonably close to the true values. Figure 6.11d, finally, shows the data together with the best nonlinear fit and the confidence levels $\pm\hat{\sigma}_n$ and demonstrates, that the NPM has achieved a good description of the data.

An example for a source separation task which is spatiotemporally non-separable together with its solution by an NPM is shown in figure 6.12. The image stack shown in figure 6.12a was generated again using the three source patterns of figure 6.7a with $32 \times 32$ pixels together with two alpha functions with rise times $\tau_1 = \tau_2 = 10$ and an exponential decay function ($\tau_3 = 20$). The delays for the two alpha functions were systematically varied: The delay of $a_1$ varied linearly along the x-direction from 18 to 22 and the delay of $a_2$ varied linearly along the negative y-direction from 4 to 8. The three sources were combined with amplitudes $(-0.5, 1, 1)$ and Gaussian noise with $\sigma_n = 0.1$ to

$$\begin{aligned} x_m(\mathbf{r}) &= -0.5 \quad a_1(t_m; \tau_1 = 10, \delta_1 = 18 + 4/31 r_x)\; s_1(\mathbf{r}) \\ &+ \quad a_2(t_m; \tau_2 = 10, \delta_2 = 8 - 4/31 r_y)\; s_2(\mathbf{r}) \\ &+ \quad a_3(t_m; \tau_3 = 20) s_3(\mathbf{r}) \\ &+ \quad n_m(\mathbf{r}) \end{aligned} \tag{6.47}$$

To this image stack, a NPM with initial parameter values $\Phi_0 = (10, 20, 10, 6, 20)$ was applied pixelwise (i.e. to each pixel time series separately and under the same conditions as in figure 6.11). The resulting statistical parametric maps for the nonlinear parameters $\delta_{1,\text{Opt}}(\mathbf{r})$ (left) and $\delta_{2,\text{Opt}}(\mathbf{r})$ (right) are shown in figure 6.12b, and figure 6.12c displays the SPMs for the linear amplitude parameters $\hat{s}_1(\mathbf{r}), \hat{s}_2(\mathbf{r}), \hat{s}_3(\mathbf{r})$ (from left to right). The linear maps are very similar to the original source patterns, which demonstrates that the NPM successfully extracted the sources from the spatiotemporally non-separable example data. The non-linear SPMs for the delays figure 6.12b correctly extracted a horizontal and vertical gradient for the delay of the second and third sources, respectively. The delays were actively estimated (differed from their initial values) only at regions with a non-zero contribution of the respective source, i.e. at regions were the corresponding spatial source pattern has non-zero values. This behavior is reasonable because nonlinear parameters cannot be reliably estimated if the

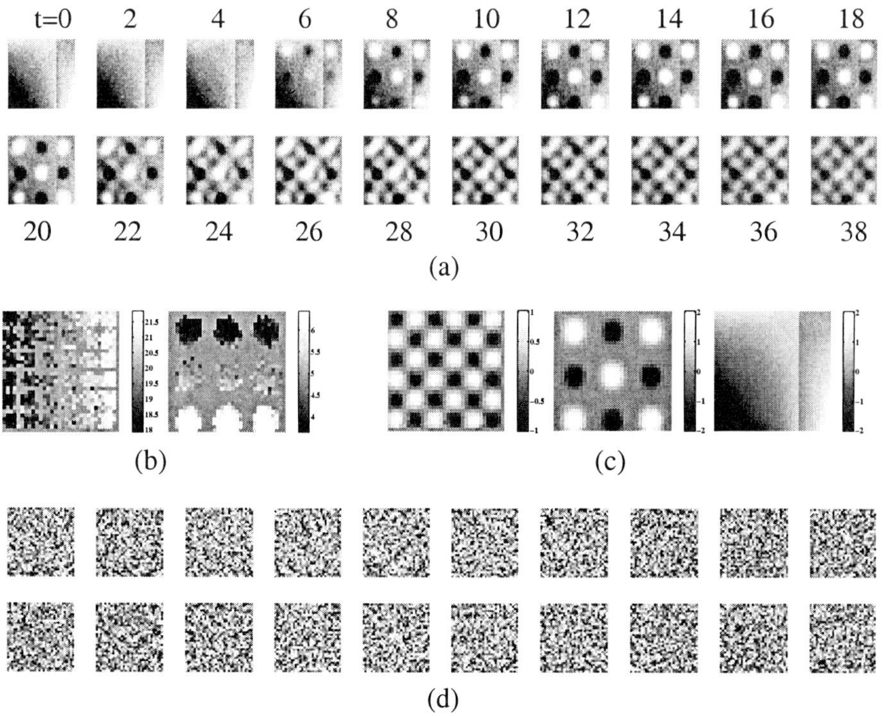


*Figure 6.12.* Source separation using a nonlinear parametric model. **(a)** artificial noisy image stack (every second image shown). Sources are non-separable in space and time because of a gradient in the delays of the alpha functions (see text). **(b)** SPM of the nonlinear delay parameters, **(c)** SPM of the linear amplitude parameters and **(d)** the residual image stack (same display as the stack in (a)). A NPM can successfully perform a source separation for the problem eq. (5.6).

corresponding model function is not significantly contained in the considered pixel time series. Figure 6.12d, finally shows the residual of the fit in which almost no structure is left.

The examples shown demonstrate the powerful potential of NPMs for the analysis of optical imaging data. However, a few limitations of this method are worth mentioning: $(i)$ NPMs represent a nonlinear optimization procedure, which can be captured in local minima. The success rate for nonlinear fits therefore depends on the initial values for the parameters. In practice, reasonable initial values for the curve shape parameters of model functions can be obtained by analyzing mean-time series of different regions of interest in a preprocessing step and obtaining initial values for the nonlinear functions from a fit to these mean-time series. Often, the initial parameters are close to the average over the nonlinear SPM, and the nonlinear parametric model only has to find the deviation of the actual parameter from the mean value. Actually, the initialization of the parameters to the mean of the true values has been used for the fit figure 6.12. $(ii)$ If the families of model functions chosen are similar to each other (as the two alpha functions in the examples), or if they do not change very strongly depending on their parameters (as the exponential function in the examples), the resulting cost function for the optimization process can show almost flat regions. In the examples above, the cost function becomes flat in the $s_1$ and $s_2$ directions, if the two alpha functions have the same non-linear parameters and opposite amplitudes and becomes flat along the $\tau_3$ direction of the exponential for large values of $\tau_3$. Part of this problem can be overcome heuristically by putting constraints on the allowed ranges of different parameters. $(iii)$ a NPM has to estimate a considerable number of parameters per pixel time series. As a consequence, this methods requires sufficiently long image stacks, which are not always available from real imaging experiments. Also, its sensitivity to noise increases compared to a GLM (cf. errors in amplitude parameters for GLM and NPM above), because the NPM has to estimate more free parameters from the data. $(iv)$ A NPM can only be applied successfully, if the true time series are located within the search space of the model families: we still have to guess the correct model family, and if we fail to do so the NPM will fail.

In the next chapter we will describe an alternative set of methods for source separation, namely projection methods, which do not rely on explicit knowledge about the time courses of signals, but instead use assumptions about their statistical structure over space and time to find interesting directions in the data space.

Chapter 7

# PROJECTION METHODS FOR SOURCE SEPARATION

## 1. Principle

Projection methods represent a family of particular algorithms, which attempt to find and represent interesting statistical structure in the data. The data to be analyzed are a set of $P$ data vectors $\mathbf{x}(\mathbf{r}) = (x_1, ..., x_M)^T(\mathbf{r})$, which are indexed by $\mathbf{r}$. Each data vector can be represented as a point within a $M$-dimensional vector space, which is called the *data space* or *state space*, and all $P$ data vectors together form a data cloud within the state space. Generally, the data points can be considered as random samples, which are drawn from a usually unknown pdf $p(\mathbf{x})$. Projection methods attempt to characterize important aspects of the underlying pdf $p(\mathbf{x})$ on the basis of the observed data set. They attempt to find a direction or a set of directions within the data space, along which particular aspects of the statistical structure of the data become obvious. In the simplest case, the projection of the data onto these directions reveal an aspect of its statistical structure. Figure 7.1 schematically illustrates two examples for interesting directions in the data space, namely a direction along which the variance of the data becomes maximal (figure 7.1a) and a direction along which the data show maximal segregation into clusters (figure 7.1a).

How can projection methods contribute to the solution of the source separation problem eq. (5.10)? In the previous chapter we have seen that regression methods for the analysis of optical imaging data may fail to provide a correct solution, if some explicit knowledge about the time courses of the different signal components is not available, or if we use the wrong search space for the time courses. As an alternative, we may consider methods, which are based on some implicit knowledge, namely on certain assumptions about the statistical structure of the data. For this approach it is useful to think of a given observed

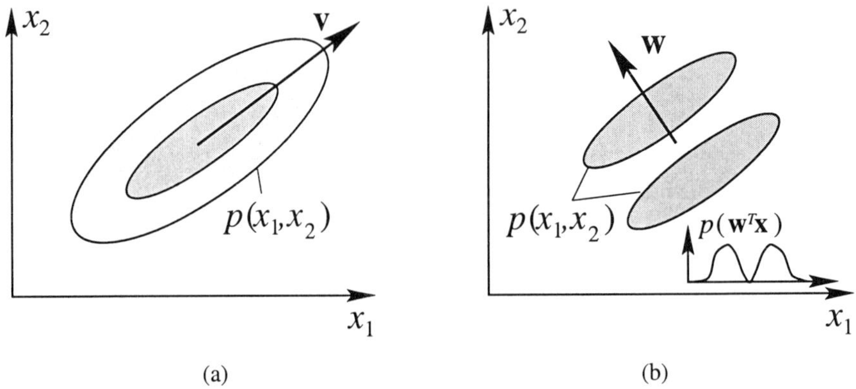


*Figure 7.1.* Illustration of the principle of projection methods. The goal is to find directions within the data space, which reveal interesting statistical structure of the underlying probability density function $p(\mathbf{x})$. The projection of the data onto the corresponding vectors $\mathbf{v}$ and $\mathbf{w}$ makes one aspect of this structure explicit. There are many interesting features including **(a)** maximal variance of the data or **(b)** maximal clustering of data.

pixel time series $\mathbf{x}$ of an optical imaging experiment as a random sample, which is drawn from the unknown pdf $p(\mathbf{x})$. Now, instead of explicitly specifying the properties of the time series itself, we assume that we know something about $p(\mathbf{x})$ – in other words we setup a generative model of the underlying probability density function.

The generative model for the pixel time series follows in a natural way from eq. (5.10): We assume that the pixel time series result as a linear mixture of source vectors plus additive noise,

$$\mathbf{x} = \mathbf{A}\,\mathbf{s} + \mathbf{n}. \tag{7.1}$$

This generative model also specifies a relationship between the pdf's of the observed mixtures and the unknown sources. As discussed in the previous chapter, our task consists of estimating the sources $\mathbf{s}$. Here we have to estimate them only on the basis of the observed data and some knowledge or assumptions about the underlying pdf's, but both the mixing matrix $\mathbf{A}$ and the sources $\mathbf{s}$ are unknown. Therefore, the present problem represents an example of a blind source separation (BSS) problem (Jutten and Herault, 1986). BSS attempts to find an estimate $\mathbf{B} = \hat{\mathbf{A}}^{-1}$ for the inverse of the mixing matrix[1]. The columns of $\mathbf{A}$ represent a particular set of projection directions, namely the desired time courses of the signal components. However, there exists an ambiguity between

[1]The true inverse only exists if $\mathbf{A}$ is a square matrix, $M = L$, and has full rank. We will henceforth restrict ourselves to this case.

the length of the column vectors of $\mathbf{A}$ (i.e. the amplitudes of the time courses $a_l(t)$) and the variance of the source patterns:

$$x_m = \sum_l a_{m,l}\, s_l = \sum_l a_{m,l}\, (\lambda_l s_l') = \sum_l (a_{m,l}\lambda_l)\, s_l' \equiv \sum_l a_{m,l}'\, s_l'. \quad (7.2)$$

In the last chapter this ambiguity was resolved by keeping the functions $a_l(t)$ normalized to unity. In the present chapter, we define the source patterns $s_l(\mathbf{r})$ to have unit variance for reasons discussed below.

Principal Component Analysis (PCA) (Bishop, 1995) removes second order correlations from the data and represents an important preprocessing tool which can already perform at least a partial separation. Full BSS algorithms often proceed from the results of PCA and use further statistical source properties in order to complete the separation. One class of BSS algorithms, which is known as Independent Component Analysis (Jutten and Herault, 1986; Bell and Sejnowski, 1995; Amari, 1996; Cardoso, 1997; Hyvärinen et al., 2001), is based on the assumption of statistical independence of different sources. A second class of BSS algorithms is based on the assumption of vanishing cross-correlations between sources that are smooth in time (Molgedey and Schuster, 1994). In the following, we will describe some approaches towards a partial or full solution of the blind source separation problem eq. (5.10) together with some examples on artificial data and performance measures. Blind separation of sources has been successfully applied to biomedical data including functional Magnetic Resonance Imaging (fMRI) (McKeown et al., 1998), electro-encephalograpical measurements (EEG) (Makeig et al., 1996), cardiovascular signals (Vetter et al., 1999) and optical imaging (Stetter et al., 1997c; Everson et al., 1997; Otto et al., 1998; Schießl et al., 1999; Schöner et al., 2000; Schießl et al., 2000b; Stetter et al., 2000c; Vollgraf et al., 2000; Schießl et al., 2000a).

Our treatment in the following chapters is based on the following assumptions: $(i)$ The mixing matrix is a full rank square matrix, $L = M$. This can be virtually always fulfilled in practical applications because of the presence of noise, which can be treated as "spurious" sources. $(ii)$ The data vectors have zero mean: $\frac{1}{P}\sum_P \mathbf{x}(\mathbf{r}) = 0$. This assumption can be fulfilled without loss of generality by subtraction of the mean time series from the original data.

## 2. Principal Component Analysis

Principal Component Analysis finds a set of orthogonal directions within the data space, referred to as principal components (PC), along which the variance of the data assumes extremal values. Figure 7.1a shows an example of an elongated two-dimensional pdf in a gray scale plot. A set of data points drawn from this pdf show a high variance along one direction and a lower variance along the other. The vector $\mathbf{v}$ represents the first principal component of this pdf, namely the direction along which the projections of the data show maximal

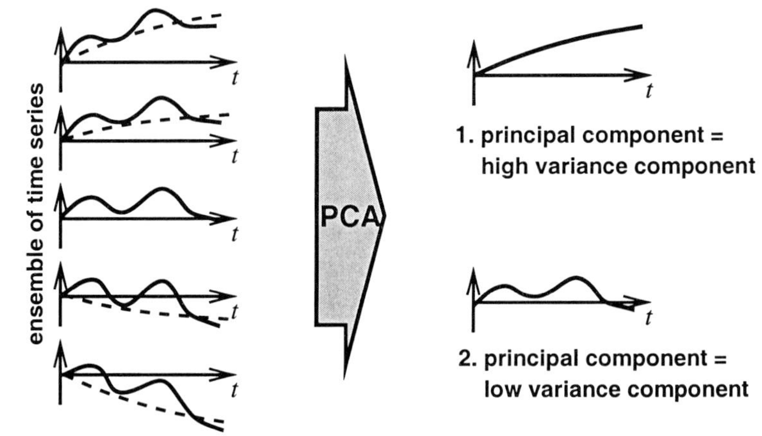


*Figure 7.2.* Illustration of PCA for time-series data (temporal PCA). PCA can separate time series the amplitudes of which vary strongly over the set of pixels (dashed lines in the ensemble of time series) from those which vary more weakly. The variance of a time series does not refer to its own amplitude within one pixel, but characterizes how strongly its amplitude changes on average over all pixels.

variance. Consequently, the projections of the data onto the first PC capture the highest portion of the variance that is present in the data. The second PC represents the direction of highest variance in the subspace orthogonal to the first PC and so forth. Because the set of all principal components is a complete orthonormal system, the whole data set can be exactly reconstructed from its projections onto all principal components. Again we identify each data point of the data set with an individual pixel time series, and the whole data set with the set of all pixel time series of an image stack. Figure 7.2 illustrates how PCA acts if each data point is a time-series. PCA can separate prototype time series (the principal components), the amplitude of which vary strongly as one proceeds through all pixels, from those with weaker variance.

## 2.1 Mathematical Formulation

The set of $M$ principal time series $\mathbf{v}_m, m = 1., ..., M$ within the data space and the data variances along these directions can be found as the solutions of the eigenvalue problem

$$\mathbf{C_x V} = \mathbf{V\Lambda}. \tag{7.3}$$

The matrix $\mathbf{C_x}$ is the covariance matrix of the data with elements

$$C_{x,mn} = \langle x_m(\mathbf{r}) x_n(\mathbf{r}) \rangle_{\mathbf{r}} := \frac{1}{P} \sum_{\mathbf{r}} x_m(\mathbf{r}) x_n(\mathbf{r}) \tag{7.4}$$

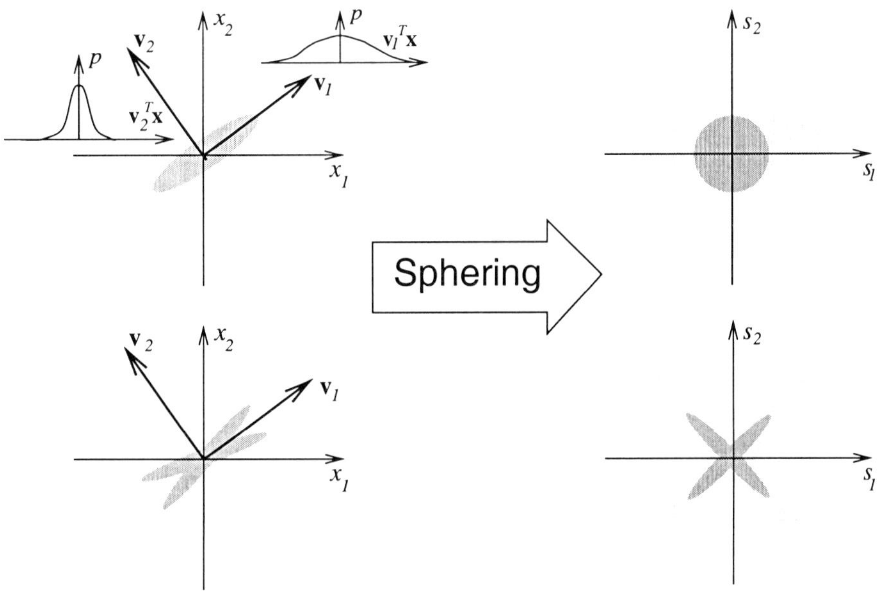


*Figure 7.3.* Illustration of the sphering procedure for a Gaussian pdf (top) and a non-Gaussian pdf (bottom). Each pdf is treated as a Gaussian, and the principal components are calculated. Then the data are transformed to the set of PCs and their variances within the new coordinate system are equalized.

and can be written in matrix notation as

$$\mathbf{C_x} = \frac{1}{P}\mathbf{X}\mathbf{X}^T \tag{7.5}$$

The orthogonal $M \times M$ matrix $\mathbf{V}$ (i.e., $\mathbf{V}^{-1} = \mathbf{V}^T$) contains the eigenvectors $\mathbf{v}_m$ as columns. $\mathbf{\Lambda}$ is the diagonal matrix with the corresponding eigenvalues $\lambda_m$ as diagonal elements. The $m$-th eigenvalue $\lambda_m$ represents the variance of the data along the $m$-th principal axis $\mathbf{v}_m$.

## 2.2 Sphering and Dimension Reduction for Optical Imaging

By use of the matrices $\mathbf{V}$ and $\mathbf{\Lambda}$ we can transform the original data in a way that the new data set is uncorrelated: the components of the transformed vectors have unit variance and all cross-correlations between them vanish. We define

$$\mathbf{Y} := \mathbf{\Lambda}^{-1/2}\mathbf{V}^T\mathbf{X} =: \mathbf{D}\mathbf{X}, \tag{7.6}$$

for which we find the correlation matrix

source separation, may not be orthogonal anymore, and cannot be found using the Jacobi method.

For noisy BSS tasks, it seems more appropriate to consider a cost function for the matrix $\mathbf{U}$, which becomes minimal, if the conditions eq. (7.34) is optimally fulfilled. But now we do not require the matrix to be orthogonal anymore, and possible deviations from orthogonality can help to correct for possible errors due to wrong sphering. We can score a separating matrix by its ability to remove the cross-correlation functions between the source estimates while keeping the autocorrelation functions finite. One possible cost function which achieves this task is

$$
\begin{aligned}
E(\mathbf{U}) &= \sum_{\{\Delta\mathbf{r}\}} \sum_{l \neq m} \langle \hat{\mathbf{s}}_m(\mathbf{r}) \hat{\mathbf{s}}_l(\mathbf{r} + \Delta\mathbf{r}) \rangle_{\mathbf{r}} && (7.36) \\
&= \sum_{\{\Delta\mathbf{r}\}} \sum_{l \neq m} \left( \left( \mathbf{U}\mathbf{C}_{\mathbf{y}}(\Delta\mathbf{r})\mathbf{U}^T \right)_{lm} \right)^2 . && (7.37)
\end{aligned}
$$

Alternatively, we can minimize a cost function, which scores the full separating matrix,

$$
E(\mathbf{B}) = \sum_{\{\Delta\mathbf{r}\}} \sum_{l \neq m} \left( \left( \mathbf{B}\mathbf{C}_{\mathbf{x}}(\Delta\mathbf{r})\mathbf{B}^T \right)_{lm} \right)^2 . \tag{7.38}
$$

Its optimization implicitly estimates both the sphering matrix $\mathbf{D}$ and the matrix $\mathbf{U}$ without any further assumptions, but the optimization can be numerically unstable, if the mixing matrix is far from orthogonal.

## 4.3 Noise-Robust Sphering

As we have seen, noise can deteriorate the zero-shift covariance matrix and can cause erroneous sphering. If the sensor noise is white in space and time, i.e., if it has a diagonal correlation matrix, then the sphering matrix can be estimated more reliably by using a symmetrized shifted correlation matrix for a small shift vector $\Delta\mathbf{r}_n$, instead of the zero-shift covariance matrix:

$$
\begin{aligned}
\hat{\mathbf{C}}_{\mathbf{x}} &= \mathbf{A}\mathbf{C}_{\mathbf{s}}(\Delta\mathbf{r}_n)\mathbf{A}^T + \langle \mathbf{n}(\mathbf{r})\mathbf{n}^T(\mathbf{r} + \Delta\mathbf{r}_n) \rangle_{\mathbf{r}} \\
&= \mathbf{A}\mathbf{C}_{\mathbf{s}}(\Delta\mathbf{r}_n)\mathbf{A}^T \approx \mathbf{A}\mathbf{C}_{\mathbf{s}}(0)\mathbf{A}^T && (7.39)
\end{aligned}
$$

## 4.4 A Unifying Approach: Convolutive Decorrelation Procedures

Single-shift and multishift ESD represent heuristic approaches towards second-order decorrelation, and can be viewed as special cases of a more general approach towards decorrelation methods (Vollgraf et al., 2000), which is based on convolved versions of the signals considered. Starting from a general formulation of convolutive decorrelation methods we derive criteria for the choice of the cost function and the pairs of convolutions for optimal performance.

### 4.4.1 Principle of Convolutive Decorrelation Methods

We wish to optimize a strategy for estimating the unmixing matrix $\mathbf{B}$ from a set of shifted covariance functions

$$\mathbf{B}\mathbf{C}_{\mathbf{x}}(\Delta\mathbf{r})\mathbf{B}^T = \mathbf{\Lambda}(\Delta\mathbf{r}), \quad , n = 1, ..., N. \tag{7.40}$$

As we shall see shortly, signals of finite length, in general, have only approximately vanishing cross correlation functions. Therefore, the choice of the shift parameters $\Delta\mathbf{r}$ is crucial for the success of the algorithm. Optimal $\Delta\mathbf{r}$ would yield minimal cross correlations in the true sources, compared to their autocorrelations. Since the sources and hence optimal $\Delta\mathbf{r}$ are unknown, we have to find strategies which minimize finite size effects and noise effects.

One general approach towards the solution of eq. (7.40) is to consider filtered or convolved versions of the signals instead of shifted ones. For discrete signals, the convolution of a signal $x_m$ with a kernel $f$ is given by

$$(x_m * f)(\mathbf{r}) = \sum_{\Delta\mathbf{r}} x_m(\mathbf{r} - \Delta\mathbf{r}) f(\Delta\mathbf{r}). \tag{7.41}$$

With $f(\Delta\mathbf{r}) = \delta(\Delta\mathbf{r} - \Delta\mathbf{r}_0)$, a shift of a spatial pattern by $\Delta\mathbf{r}_0$ results as a special case of a convolution. A unified approach to BSS based on decorrelation uses correlation matrices of a set of convolved versions of mixtures,

$$\mathbf{C}_{\mathbf{x}}^{f,g} = \langle (\mathbf{x} * f)(\mathbf{x} * g)^T \rangle_{\mathbf{r}}, \tag{7.42}$$

in order to estimate the sources. $(\mathbf{x}*f)$ denotes the component-wise convolution of the signal vector $\mathbf{x}$ with $f$. A convolutive covariance matrix $\mathbf{C}_{\mathbf{x}}^{f,g}$ can be viewed as a linear combination of all shifted covariance matrices, weighted with the values of the convolution kernels $f$ and $g$ at the corresponding shifts

$$\mathbf{C}_{\mathbf{x}}^{f,g} = \sum_{\Delta\mathbf{r}_1} \sum_{\Delta\mathbf{r}_2} f(\Delta\mathbf{r}_1) g(\Delta r_2) \mathbf{C}_{\mathbf{x}}(\Delta\mathbf{r}_2 - \Delta\mathbf{r}_1). \tag{7.43}$$

$\mathbf{C}_{\mathbf{x}}^{f,g}$ can only be diagonal, independently of $f$ and $g$, if $\mathbf{C}_x(\Delta\mathbf{r})$ is diagonal for all $\Delta\mathbf{r}$. Conversely, if all $\mathbf{C}_{\mathbf{x}}(\Delta\mathbf{r})$ are diagonal, then the same is true for $\mathbf{C}_{\mathbf{x}}^{f,g}$. We conclude that signals have zero cross correlation functions if, and only if, arbitrary convolved or filtered versions of the signals are uncorrelated at zero time delay. Consequently, the task of source separation can be carried out by simultaneously diagonalizing convolutive instead of shifted covariance matrices. We arrive at the following set of conditions for the unmixing matrix $\mathbf{B}$:

$$\mathbf{C}_{\hat{\mathbf{s}}}^{f,g} = \mathbf{B}\mathbf{C}_{\mathbf{x}}^{f,g}\mathbf{B}^T = \mathbf{\Lambda}^{(f,g)} \tag{7.44}$$

for all considered pairs of convolution kernels $(f, g)$. $\mathbf{C}_{\hat{\mathbf{s}}}^{f,g} = \langle (\hat{\mathbf{s}} * f)(\hat{\mathbf{s}} * g)^T \rangle_{\mathbf{r}}$ denotes the convolutive correlation matrix of the estimated sources, which we wish to bring as close as possible to a diagonal matrix $\mathbf{\Lambda}^{(f,g)}$. The single-shift and multishift ESD algorithms are included in the approach eq. (7.44) as special cases with shift operations as convolutions and their characterization is included in the following analysis of the general problem eq. (7.44).

### 4.4.2 Finite Signals

The effects of finite signal size onto the BSS task can be well analyzed in the frequency domain. We denote the Fourier transform of a signal $s$ by the respective upper case letter $S$. For infinite, discrete signals $s_m$, the Fourier transform is given by

$$S_m(\mathbf{k}) = \sum_{\mathbf{r}} s_m(\mathbf{r}) e^{(-i\mathbf{kr})}, \tag{7.45}$$

where $i$ is the imaginary unit and $\mathbf{k}$ denotes a vector of spatial frequencies (wave vector) in the two-dimensional space. The transformation to the frequency domain allows the cross correlation functions of the sources, $\mathbf{C_s}(\Delta\mathbf{r})$, to be easily computed by a conjugate complex multiplication. In the frequency domain, the assumption of zero cross-correlation functions eq. (7.26) is given by

$$S_m(\mathbf{k}) S_l^*(\mathbf{k}) = 0 \qquad \text{for} \quad m \neq l, \qquad \text{and any} \quad \mathbf{k}, \tag{7.46}$$

where the superscript * denotes the conjugate complex of a number. Condition eq. (7.46) says that the frequency components of uncorrelated signals must be sparse, i.e. for any spatial frequency $\mathbf{k}$, no more than one signal may be active.

A signal of finite length can be considered as an infinite signal, multiplied with a window function, which is different from zero only for a range of finite length. The multiplication in the spatial domain becomes a convolution in the spatial frequency domain, and the spectra of the signals are blurred with the spectrum of the window function. Equation (7.46) therefore holds only approximately and the cross correlation functions do not vanish completely anymore.

### 4.4.3 Symmetry of Matrices

We want to diagonalize correlation matrices such that the resulting separating matrix and the source estimates are real functions, but shifted or convolutive covariance matrices are not necessarily symmetric. Consequently, they can have complex eigenvectors and eigenvalues. We can analyze the diagonalization of an arbitrary matrix $\mathbf{C}$ by considering its decomposition into a symmetric and an antisymmetric part:

$$\mathbf{C} = \mathbf{C}^s + \mathbf{C}^a, \quad \mathbf{C}^s = \frac{1}{2}(\mathbf{C} + \mathbf{C}^T), \quad \mathbf{C}^a = \frac{1}{2}(\mathbf{C} - \mathbf{C}^T). \tag{7.47}$$

In order to successfully diagonalize $\mathbf{C}$, we have to fulfill the condition

$$\mathbf{B}\mathbf{C}\mathbf{B}^T = \mathbf{B}\mathbf{C}^s\mathbf{B}^T + \mathbf{B}\mathbf{C}^a\mathbf{B}^T = \mathbf{\Lambda} \tag{7.48}$$

where $\mathbf{\Lambda}$ is diagonal and therefore symmetric. However, real separating matrices $\mathbf{B}$ do not change the symmetry of the corresponding matrix, i.e. we find

$$\begin{aligned}\left(\mathbf{B}\mathbf{C}^{s/a}\mathbf{B}^T\right)_{lm} &= \sum_{i,j} b_{li} c_{ij}^{s/a} b_{mi} \\ &= \sum_{j,i} b_{mj} (\pm c_{ji}^{s/a}) b_{li} \\ &= \pm \left(\mathbf{B}\mathbf{C}^{s/a}\mathbf{B}^T\right)_{ml},\end{aligned} \tag{7.49}$$

where the positive sign corresponds to the symmetric and the negative sign to the antisymmetric matrix. If we know that the unmixing matrix $\mathbf{B}$ must be real, we can conclude that the antisymmetric part in eq. (7.48) must vanish. Any finite antisymmetric contribution of covariance matrices in the observed data must be due to influences of noise or finite signal size. Additionally, we cannot use the antisymmetric part of the covariance matrix in order to correct errors within the symmetric part, because both matrix spaces are orthogonal to each other. Due to these reasons, we can reduce most of the noise and finite size effects if we omit the antisymmetric part and use only symmetrized covariance matrices.

### 4.4.4 Cost Functions

It is important to know, which properties of cost functions are advantageous for the efficient simultaneous diagonalization of the equations (7.44). For diagonalization of more than two covariance matrices, an analytic solution like in (Molgedey and Schuster, 1994) does not exist anymore. Further, since the sources are usually only approximately uncorrelated, the best achievable solution would still not completely diagonalize the covariance matrices. Therefore, a measure is needed, which evaluates, how close the covariance matrices of the reconstructed signals $\mathbf{C}_{\hat{\mathbf{s}}}^{f,g} = \mathbf{B}\mathbf{C}_{\mathbf{x}}^{f,g}\mathbf{B}^T$ come to a diagonal form. This measure is implemented by a cost function, which has to be minimized with respect to $\mathbf{B}$.

One simple choice for a cost function penalizes the squared sum over all off-diagonal elements of $\mathbf{C}_{\hat{\mathbf{s}}}^{f,g}$.

$$E_1(\mathbf{B}) = \sum_{\{(f,g)\}} \sum_{l} \sum_{m \neq l} \left(\mathbf{B}\mathbf{C}_{\mathbf{x}}^{f,g}\mathbf{B}^T\right)_{lm}^2, \tag{7.50}$$

where the first sum runs over all pairs of convolution kernels $(f, g)$. The gradient of (7.50) has the form

$$\nabla E_1(\mathbf{B}) = 4 \sum_{\{(f,g)\}} \left(\mathbf{C}_{\hat{\mathbf{s}}}^{f,g} - diag(\mathbf{C}_{\hat{\mathbf{s}}}^{f,g})\right) \mathbf{B}\mathbf{C}_{\mathbf{x}}^{f,g}. \tag{7.51}$$

It can be seen, that the gradient vanishes if all $\mathbf{C}_{\hat{\mathbf{s}}}^{f,g}$ are diagonal, but the gradient also vanishes, when $\mathbf{B}$ identically vanishes, which would be the absolute minimum of the cost function in the case of not ideally uncorrelated sources. It is therefore necessary to either constrain $\mathbf{B}$ explicitly in order to prevent it from growing to small, or to setup a cost function, which is invariant against changes in the variance of the outputs $\hat{\mathbf{s}}$.

In order to overcome this problem, it has been suggested (Molgedey and Schuster, 1994) to constrain the diagonal elements of the estimated mixing matrix to a fixed value, i.e.,

$$E_2(\mathbf{B}) = E_1(\mathbf{B}) \text{ with } \mathbf{B}_{ll}^{-1} \equiv 1, \quad l = 1, ..., M \tag{7.52}$$

during minimization of the cost function (7.50). However, this constraint does not prevent $\mathbf{B}$ from becoming arbitrary small. Consider a matrix

$$\mathbf{B}^{-1} = \begin{pmatrix} 1 & a & \cdots & a \\ a & 1 & \cdots & a \\ \vdots & \vdots & \ddots & \vdots \\ a & a & \cdots & 1 \end{pmatrix}, \tag{7.53}$$

for which (7.52) is fulfilled. For large $a$, the corresponding matrix $\mathbf{B}$ is

$$\mathbf{B} = \frac{1}{a} \begin{pmatrix} -\frac{M-2}{M-1} & \frac{1}{M-1} & \cdots & \frac{1}{M-1} \\ \frac{1}{M-1} & -\frac{M-2}{M-1} & \cdots & \frac{1}{M-1} \\ \vdots & \vdots & \ddots & \vdots \\ \frac{1}{M-1} & \frac{1}{M-1} & \cdots & -\frac{M-2}{M-1} \end{pmatrix}, \tag{7.54}$$

where $M$ is the number of sources. Increasing the value of $a$ will decrease the row norm of $\mathbf{B}$ and hence decrease the cost function without diagonalizing $\mathbf{C}_{\hat{s}}^{f,g}$.

The following numerical example illustrates the weaknesses of the constraint eq. (7.52). Consider two covariance matrices

$$\mathbf{C}_{\mathbf{s}}^{1} = \begin{pmatrix} 1 & 0 \\ 0 & 2 \end{pmatrix} \tag{7.55}$$

and

$$\mathbf{C}_{\mathbf{s}}^{2} = \begin{pmatrix} 2 & 0 \\ 0 & 1 \end{pmatrix} \tag{7.56}$$

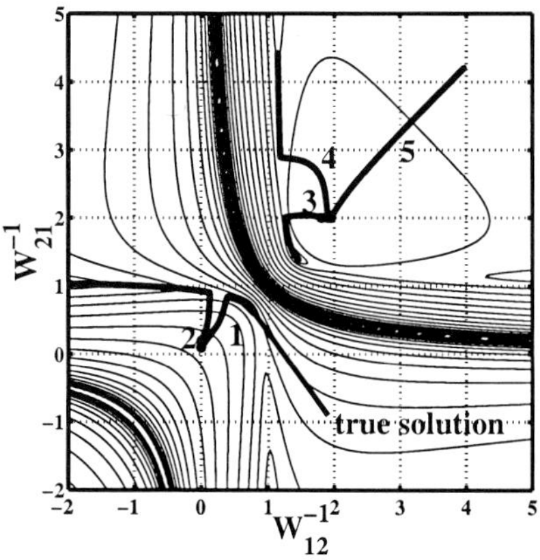


*Figure 7.8* Surface of the cost function eq. (7.52) with the constraint $\mathbf{B}_{ll}^{-1} = 1$ for two sources and the mixing matrix provided in the text. The two hyperbolic ridges are regions with singular $\mathbf{B}^{-1}$ and have infinite values of the cost function.

for two given source signals, which are mixed using the matrix

$$\mathbf{A} = \begin{pmatrix} 1 & 0.7 \\ 0.7 & 1 \end{pmatrix}. \tag{7.57}$$

Figure 7.8 shows a contour plot of the cost function with respect to the two off-diagonal elements of the estimated mixing matrix $\mathbf{B}^{-1}$. The hyperbolic ridge corresponds to values at which $\mathbf{B}^{-1}$ becomes singular and causes an infinite value of the cost function. Solid lines mark gradient descent trajectories for 5 different initializations. It can be seen, that depending on the initialization, the gradient descent procedure may succeed to find the true mixing matrix (trajectories 1,3), or may diverge, leading to arbitrary small $\mathbf{B}$ and hence trivial minima (trajectories 2,4,5). Due to these properties, the constraint (7.52) does not seem to be suitable for gradient based joint diagonalization.

The spurious minima of the simple cost function (7.50) arise from vanishing variances of the output signals $\hat{\mathbf{s}}$. To avoid this behavior, it seems reasonable to constrain the variances of the source estimates, which corresponds to constraining the unmixing matrix instead of the mixing matrix. The source variance can be constrained by normalizing the off-diagonal elements of $\mathbf{C}_{\hat{\mathbf{s}}}^{f,g}$ by the diagonal elements. Then, the gradient of the cost function is zero for directions, which only change the source variance by changing the row norm of $\mathbf{B}$. The modified cost function is given by

$$E_3(\mathbf{B}) = \sum_{\{(f,g)\}} \sum_l \sum_{m \neq l} \frac{\left(\mathbf{C}_{\hat{\mathbf{s}}}^{f,g}\right)_{lm}^2}{\left|\left(\mathbf{C}_{\hat{\mathbf{s}}}^{f,g}\right)_{ll}\left(\mathbf{C}_{\hat{\mathbf{s}}}^{f,g}\right)_{mm}\right|}. \tag{7.58}$$

Because the matrices $\mathbf{C}_{\hat{\mathbf{s}}}^{f,g}$ are not necessarily positive definite, the normalizing term must appear as absolute value in the denominator of (7.58). The gradient

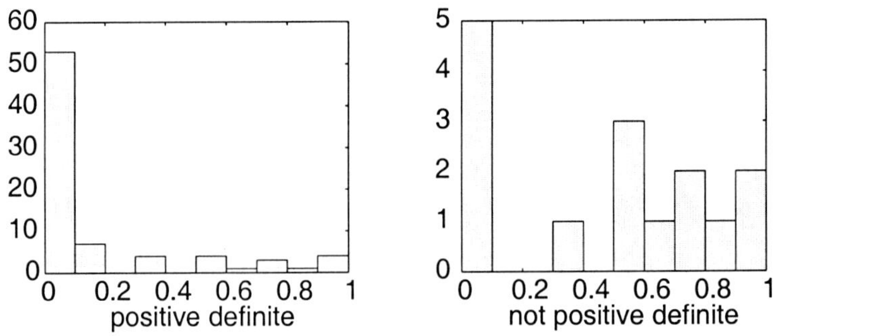


*Figure 7.9.* Histogram of the reconstruction error of the successful trials (finite reconstruction error) with positive definite $\mathbf{C}_{\hat{\mathbf{s}}}^{f,g}$ (left) and not positive definite $\mathbf{C}_{\hat{\mathbf{s}}}^{f,g}$ (right). Positive definite covariance matrices perform considerably better.

can be obtained from

$$\frac{\partial E_3}{\partial \mathbf{B}_{lm}} = 2 \sum_{\{(f,g)\}} \sum_{j \neq l} \frac{\mathbf{C}_{lj}}{|\mathbf{C}_{ll}\mathbf{C}_{jj}|} \left( \left( \mathbf{B}\mathbf{C}_{\mathbf{x}}^{f,g} \right)_{jm} - 2 \frac{\mathbf{C}_{lj}}{\mathbf{C}_{ll}} \left( \mathbf{B}\mathbf{C}_{\mathbf{x}}^{f,g} \right)_{lm} \right), \tag{7.59}$$

where $\mathbf{C}_{lm}$ abbreviates $\left(\mathbf{C}_{\hat{\mathbf{s}}}^{f,g}\right)_{lm}$.

Simulations have shown, that this cost function performs well as long as all covariance matrices are positive definite. Two experiments with 80 simultaneous diagonalizations each were performed. Each experiment consisted of the simultaneous diagonalization of four $4 \times 4$ covariance matrices. For the first 80 experiments, only positive definite matrices with random eigenvalues in $\left(\mathbf{C}_{\hat{\mathbf{s}}}^{f,g}\right)_{ll} \in ]0,1]$ were used, whereas in the second 80 trials matrices were allowed to contain entries of arbitrary sign: $\left(\mathbf{C}_{\hat{\mathbf{s}}}^{f,g}\right)_{ll} \in [-1,1]$. For each experiment, the mixing matrix $\mathbf{A}$ was initialized with elements from a (0,1)-normal distribution. The results were evaluated, using the reconstruction error $RE(\mathbf{BA})$. This quantity provides a measure of the ratio of the off-permutation elements of $\mathbf{BA}$ to the permutation elements. The permutation is determined by the largest elements in each row. If $\mathbf{BA}$ could not be interpreted as a permutation matrix, i.e. two or more maximum elements in one row or column, the source separation was considered as failed. Otherwise the reconstruction error was scored using the performance index as suggested by (Koehler and Orglmeister, 1999):

$$RE(\mathbf{G}) = \frac{1}{M} \sum_{i=1}^{M} \frac{1}{M-1} \left( \sum_{j=1}^{M} \frac{|G_{ij}|}{max_k |G_{ik}|} - 1 \right). \tag{7.60}$$

In the experiment with the positive definite $\mathbf{C}_{\mathbf{s}}^{f,g}$, a finite reconstruction error could be achieved in 77 of 80 trials, whereas with the not positive definite $\mathbf{C}_{\mathbf{s}}^{f,g}$ only 15 of 80 separations where successful. Figure 7.9 shows the histogram of the reconstruction error of the successful trials. It is on average much lower if all matrices $\mathbf{C}_{\mathbf{s}}^{f,g}$ are positive definite.

Another cost function that is independent of the output variances compares the diagonal elements of the matrices $\mathbf{C}_{\hat{s}}^{f,g}$ with their eigenvalues:

$$E_4(\mathbf{B}) = \sum_{\{(f,g)\}} \mathcal{D}\left(diag\left(\mathbf{C}_{\hat{\mathbf{s}}}^{f,g}\right), eig\left(\mathbf{C}_{\hat{\mathbf{s}}}^{f,g}\right)\right) \tag{7.61}$$

The operator $diag$ extracts the vector of diagonal elements of a matrix and the operator $eig$ the vector of eigenvalues. $\mathcal{D}$ is a distance measure between the vector of the eigenvalues and the vector of diagonal elements of $\mathbf{C}_{\hat{\mathbf{s}}}^{f,g}$. For diagonal $\mathbf{C}_{\hat{\mathbf{s}}}^{f,g}$ these two vectors are identical (provided they are properly sorted by magnitude). To obtain invariance on output variances, the diagonal elements need to be normalized to unity

$$\left(\tilde{\mathbf{C}}_{\hat{\mathbf{s}}}^{f,g}\right)_{lm} = \left(\mathbf{C}_{\hat{s}}^{f,g}\right)_{ml} \left|\left(\mathbf{C}_{\hat{\mathbf{s}}}^{f,g}\right)_{ll} \left(\mathbf{C}_{\hat{\mathbf{s}}}^{f,g}\right)_{mm}\right|^{-\frac{1}{2}}. \tag{7.62}$$

Thus, the cost function has the form

$$E_4(\mathbf{B}) = \sum_{\{(f,g)\}} \left(\sum_l \left|1 - |\lambda_l^{f,g}|\right|^{\alpha_1}\right)^{\alpha_2}, \tag{7.63}$$

where the parameters $\alpha_1$ and $\alpha_2$ control the distance metric and $\lambda_l^{f,g}$ are the eigenvalues of $\tilde{\mathbf{C}}_{\hat{s}}^{f,g}$.

This cost function performs quite similar to eq. (7.58) as shown in figure 7.10, where for 2000 separating matrices near the absolute minimum $\mathbf{A}^{-1}$ of the experiment of figure 7.8, the values of both cost functions are plotted against each other. The narrow shape of the cloud corresponds to a strong correlation between the two cost functions, but eq. (7.63) is quite computationally expensive due to the need of numerically calculating the gradient, which includes multiple evaluations the eigensystem of $\mathbf{C}_{\hat{\mathbf{s}}}^{f,g}$.

In summary, the cost function (7.58) seems to be most advantageous, because it is fast to compute and for positive $\mathbf{C}_{\hat{\mathbf{s}}}^{f,g}$ avoids spurious minima corresponding to trivial solutions.

### 4.4.5 Choice of Convolution Kernels

For convolutive decorrelation algorithms we may freely choose the set of convolutions $f$ and $g$, which we want to use for source separation. The following considerations provide two criteria for good choices of convolution kernels.

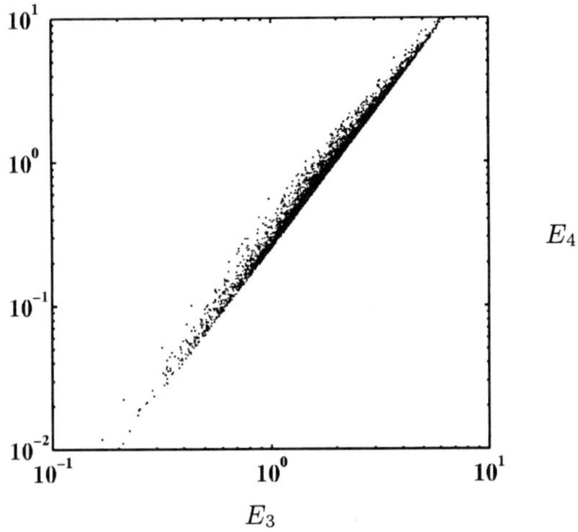


*Figure 7.10.* Logarithmic plot of the values of the cost functions $E_3$ and $E_4$ for the example eqs. (7.55 - (7.57) and 2000 separating matrices near the global minimum.

The first criterion is based on the maximization of noise robustness. In the presence of white noise, a noise vector is added to the linear mixture, $\mathbf{x} = \mathbf{A}\mathbf{s} + \mathbf{n}$. The noise signals are considered to have zero cross correlation functions with each other and with the signals, i.e. $(\mathbf{C_n}(\Delta\mathbf{r}))_{ml} = 0$ for all $\Delta\mathbf{r}$ and all $l \neq m$. Thus, the best source separation result for $\mathbf{B} = \mathbf{A}^{-1}$ yields

$$\hat{\mathbf{s}} = \mathbf{B}\mathbf{x} = \mathbf{s} + \mathbf{A}^{-1}\mathbf{n}. \tag{7.64}$$

That is, in the best case, the sources are properly separated, but the noise is still present. In the general case, the covariance matrices of the outputs can be written as

$$\mathbf{C}_{\hat{\mathbf{s}}}^{f,g} = \mathbf{B}\mathbf{A}\mathbf{s}\mathbf{A}^T\mathbf{B}^T + \mathbf{B}\mathbf{C}_{\mathbf{n}}^{f,g}\mathbf{B}^T. \tag{7.65}$$

Optimal convolution kernels for noise robustness should suppress $\mathbf{C}_{\mathbf{n}}^{f,g}$, which is, equivalently to (7.43),

$$\mathbf{C}_{\mathbf{n}}^{f,g} = \sum_{\mathbf{r}_1}\sum_{\mathbf{r}_2} f(-\mathbf{r}_1)g(-\mathbf{r}_2)\mathbf{C_n}(\mathbf{r}_2 - \mathbf{r}_1) = 0 \tag{7.66}$$

The noise is assumed to be white, thus $\mathbf{C_n}(\mathbf{r}_2 - \mathbf{r}_1) = 0$ for $\mathbf{r}_1 \neq \mathbf{r}_2$. Thus, the noise term in (7.65) is zero when $f$ and $g$ are orthogonal.

$$\sum_{\mathbf{r}} f(\mathbf{r})g(\mathbf{r}) = 0 \tag{7.67}$$

The second criterion is based on the observation (figure 7.9), that convolutive decorrelation using the cost function (7.58) work best, if the convolutive correlation functions are positive definite. Here we examine if one can generally find convolution kernels, that guarantee positive definite $\mathbf{C}_{\hat{\mathbf{s}}}^{f,g}$ and $\mathbf{C}_{\mathbf{x}}^{f,g}$. Because positive definiteness of $\mathbf{C}_{\hat{\mathbf{s}}}^{f,g}$ and $\mathbf{C}_{\mathbf{x}}^{f,g}$ are equivalent, we can test for positive definiteness on the basis of the mixtures instead of the sources. The goal is to achieve positive diagonal elements of $\mathbf{C}_{\hat{\mathbf{s}}}^{f,g}$. The $m$-th diagonal element is the correlation of the two signals $s_m * f$ and $s_m * g$. This quantity can be calculated in the frequency domain

$$\left(\mathbf{C}_{\hat{\mathbf{s}}}^{f,g}\right)_{mm} = \frac{1}{2\pi} \sum_{\mathbf{k}} F(\mathbf{k}) G^*(\mathbf{k})\, S_m(\mathbf{k}) S_m^*(\mathbf{k}). \tag{7.68}$$

Making use of the conjugate complex symmetry of the spectrum of real signals, we obtain the desired condition

$$\Re\left(F(\mathbf{k})G^*(\mathbf{k})\right) \geq 0 \qquad \text{for all} \quad \mathbf{k}, \tag{7.69}$$

where $\Re(z)$ denotes the real part of a complex number $z$. The equality must not hold for at least one spatial frequency vector $\mathbf{k}$ with $S_m(\mathbf{k}) \neq 0$, otherwise we would get a zero diagonal element in $\mathbf{C}_{\mathbf{s}}^{f,g}$ (and also in $\mathbf{C}_{\hat{\mathbf{s}}}^{f,g}$) which would lead to a singularity in the normalization term of cost function (7.58). Hence, in the general case, this finding is incompatible with (7.67), what can be seen from

$$\sum_{\mathbf{r}} f(\mathbf{r}) g(\mathbf{r}) = (F * G)|_{\mathbf{k}=0} = \sum_{\mathbf{k}} \Re\left(F(\mathbf{k})G^*(\mathbf{k})\right). \tag{7.70}$$

Equation (7.70) can only be zero if all sum terms on the right hand side are zero or negative terms are allowed.

Thus, positive definite covariance matrices cannot be generally guaranteed with orthogonal $(f, g)$. However, this does not mean, that it is impossible to obtain positive definite covariance matrices from orthogonal convolution kernels. If prior knowledge about the sources is available (e.g. if they are restricted to a certain frequency band), this knowledge can be used together with eq. (7.68) for the construction of convolution kernels, which fulfill eqs. (7.67) and (7.70) simultaneously.

## 5. Examples and Benchmarks

The success of the separation performance among different BSS algorithms strongly depends on the statistical properties of the data set (Schießl et al., 1999). Figure 7.11a shows the same set of artificial source patterns as used in the previous chapter 6. The sources have been designed as spatially smooth but otherwise arbitrary patterns. In this data set, the autocorrelation functions are smooth and are different from each other, but the cross-correlation functions

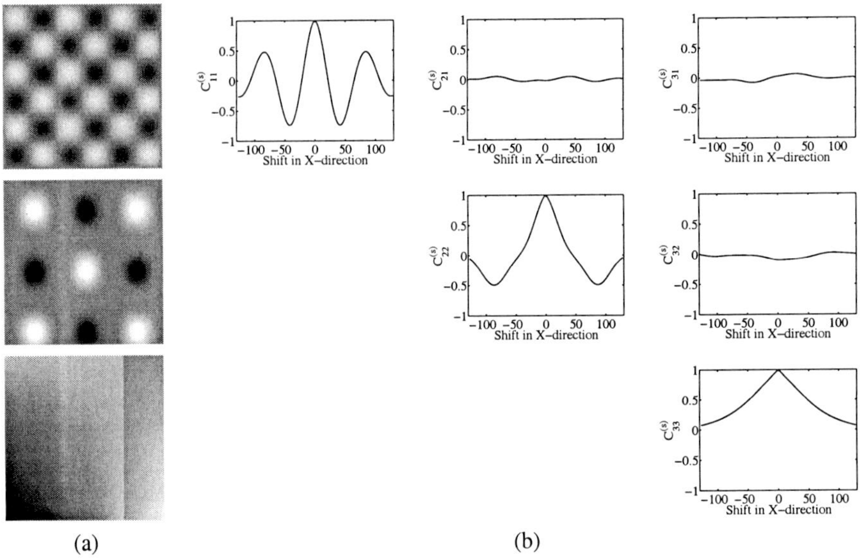


*Figure 7.11.* **(a)** Set of three artificial source patterns $s_1$ (top), $s_2$ (middle) and $s_3$ (bottom) as in figure 6.7, sampled with $256 \times 256$ pixels. **(b)** One-dimensional slices through their auto- and cross-correlation functions. The sources have smooth and different autocorrelation functions, but their cross-correlation functions vanish almost completely.

vanish almost completely. During the generation of several artificial source patterns, we found that the first property has to be designed by hand, but the second property of small cross-correlation functions emerges naturally for most patterns. In contrast, statistical independence is not a natural feature both of arbitrary artificial sources and of natural scenes (Oja, 1997). Figure 7.12 illustrates, how two-dimensional slices $p(s_1, s_2)$ (7.12a), $p(s_1, s_3)$ (7.12b) and $p(s_2, s_3)$ (7.12c) through the joint distribution of the artificial sources in figure 7.11 are shaped. The pdf is far from factorial.

In optical imaging of intrinsic signals, it is not completely clear, how far the prototype patterns fulfill the independence assumptions and to what extent their cross-correlation functions vanish. Hence, the artificial patterns of figure 7.11, which are neither completely independent nor completely uncorrelated, represent a good benchmark for the performance of different BSS algorithms.

Figure 7.13 shows an example of the performance of single-shift ESD for a noiseless mixture of the original patterns. The three sources figure 7.13a where normalized to zero mean and unit variance and mixed using the random matrix

$$\mathbf{A} = \begin{pmatrix} 0.39 & -0.56 & 0.78 \\ 0.08 & 0.44 & 0.57 \\ -0.64 & -0.95 & -0.82 \end{pmatrix} \tag{7.71}$$

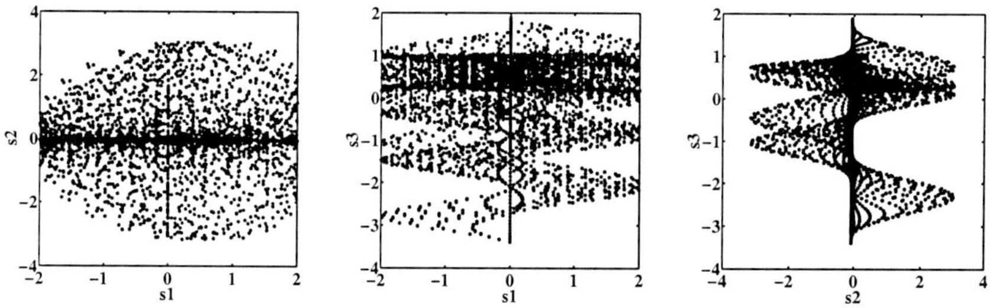


(a) (b) (c)

*Figure 7.12.* Scatter plots **(a)** of the data vectors $(s_1, s_2)$, **(b)** the vectors $(s_1, s_3)$ and **(c)** the vectors $(s_2, s_3)$. of the three artificial source patterns in figure 7.11, which approximate two-dimensional slices through the joint pdf $p(s_1, s_2, s_3)$. The source components show expressed statistical dependencies.

with a condition number of 6.5 (figure 7.13b). Figure 7.13c and 7.13d display the reconstruction of the patterns after sphering and after single-shift ESD, respectively. The BSS technique successfully reconstructs the three source patterns up to their sign and order in spite of the presence of small nonzero cross-correlations.

Because optical imaging data sets contain strong photon shot noise and are located in the regime of low signal-to-noise ratio close to 0 dB, it is also important to benchmark the robustness of different BSS algorithms against sensor noise $\mathbf{n}$. For a combined noise and statistical dependency benchmark, we generated many noisy mixtures of the sources using the matrix eq. (7.71). To each of the resulting mixtures different Gaussian white noise with variance $\sigma_n^2$ was added. ICA based on Infomax, fastICA by kurtosis maximization and single-shift ESD were then applied to the noisy mixtures resulting in the source estimates $\hat{s}_l, l = 1, 2, 3$.

The quality of reconstruction was automatically scored by calculating the covariance matrix $G_{ml} = \langle \hat{s}_m s_l \rangle_r$ between the true source and the estimates. In the ideal case, the result would be a permutation matrix, that is one component within each row and column. Note, that this criterion directly judges the sources and is different from the criterion $\mathbf{G} = \mathbf{BA}$ used in section 4.4, which is based on the matrices and measures, how well the separating matrix inverts the mixing matrix. Judging the sources is desirable in the noisy case, because the optimal unmixing matrix might differ from $\mathbf{A}^{-1}$: Even if $\mathbf{B}$ would be found to yield $\mathbf{B} = \mathbf{A}^{-1}$, the source estimates $\hat{\mathbf{s}} = \mathbf{s} + \mathbf{A}^{-1}\mathbf{n}$ still would be suboptimum, because they are corrupted by the noise. Instead, the optimal unmixing matrix can differ from $\mathbf{A}^{-1}$, because it corrects for the sensor noise as good as possible. The reconstruction error was determined as $RE(\mathbf{G})$.

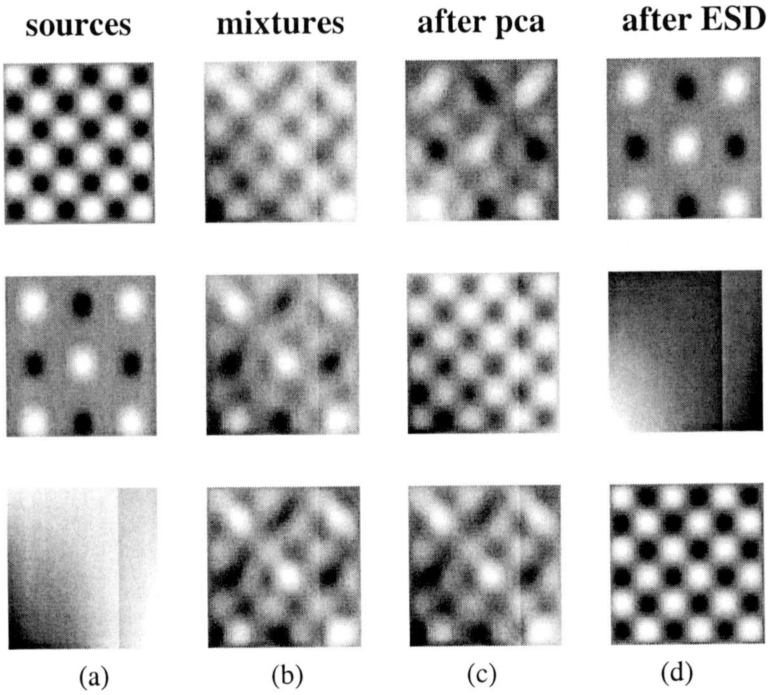


*Figure 7.13.* Reconstruction of spatially smooth signal sources **(a)** from a noiseless mixture **(b)** according to eqs. (7.1) and (7.71) using single-shift ESD. **(c)** Mixtures after sphering by PCA. The separation is incomplete. **(d)** Successful separation after full ESD with shift vector $\Delta \mathbf{r} = (5, 5)$. ESD is robust against moderately non-zero cross-correlations.

Figure 7.14 shows the dependence of the mean error on the signal to noise ratio for Infomax (figure 7.14a), fastICA (kurtosis maximization, figure 7.14b) and single-shift ESD (figure 7.14c). The signal-to-noise ratio (SNR) is the ratio between the largest variance of the three mixtures and the variance of the added noise in db. For each SNR 25 trials with different sensor noise were carried out (circles). The solid line indicates the percentage of successful separations, the dashed line marks the percentage of permutations matrixes counted in a set of 3x3 random matrices for comparison. It represents the chance level for obtaining a "successful" separation. The separation performance at the limit of high SNR is a measure for the robustness of the respective method against statistical dependencies or nonzero cross-correlations. Both ICA algorithms 7.14a and 7.14b work well in presence of statistical dependencies, but fastICA shows a lower reconstruction error. This enhanced robustness might be due to the fact, that fastICA evaluates only the kurtosis and not the full shape of the pdf. The best and most stable reconstruction, however, is achieved by ESD.

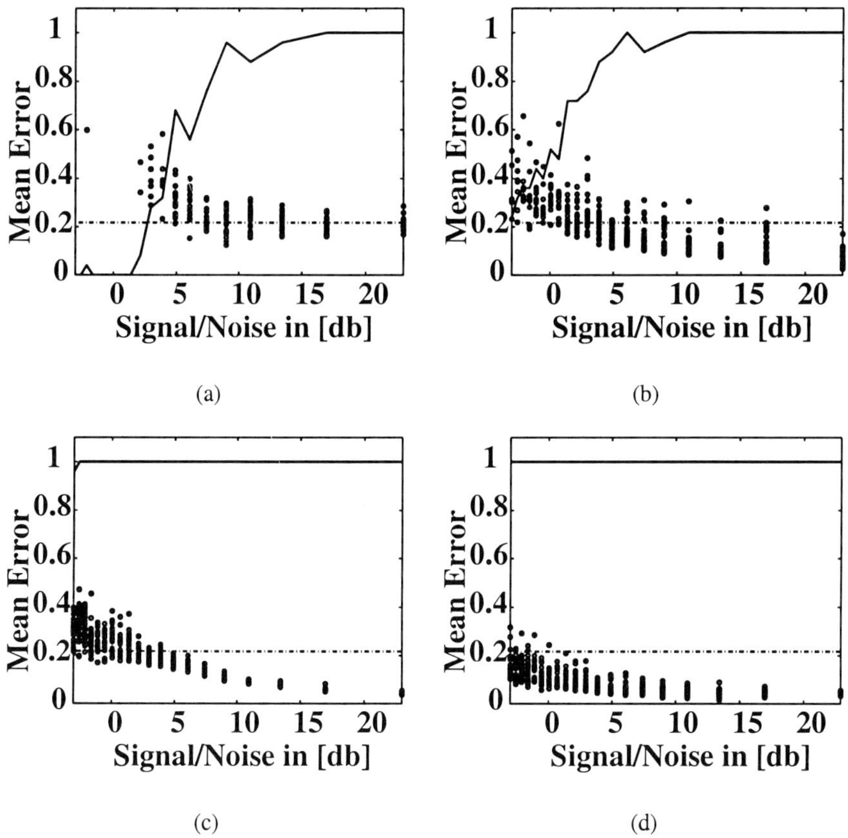


*Figure 7.14.* Reconstruction error as a function of the signal to noise ratio in db (25 trials per noise level), **(a)** Infomax, **(b)** fastICA and **(c)** extended spatial decorrelation (shift vector $\Delta\mathbf{r} = (5, 5)$ pixels). Data were the smooth sources shown in figure 7.11 (left).**(d)** Performance of ESD after lowpass filtering of the mixtures (cutoff-frequency = 25 cycles / 256 pixels). Circles: individual trials. Solid line: percentage of successful trials. Dashed line: percentage of permutation matrices by random generation of $3 \times 3$ matrices.

The reconstruction error is very small and has a low variance. This corroborates the observation of a robust source reconstruction in figure 7.13.

Noise robustness of the methods is scored by their behavior for lower SNR. In all cases, the average quality of the separation and the variance of reconstruction errors get worse with rising noise. Compared on the individual noise levels, ESD outperforms the two ICA algorithms on the model data set. In addition, the success rate of ESD stays stable at 100 percent of the trials, even for the strongly noisy case, while the performance of the ICA algorithms is much more noise sensitive – they break down at intermediate noise levels.

robust sphering using eqs (7.39),(7.3) and (7.6) (index "1"). The algorithm "dpa" finds the separating matrix by numerically minimizing the cost function eq. (7.37) of the sphered data. The minimization was carried out using the Polak Ribiere conjugate gradient technique, where the line search is substituted by a dynamic step width adaptation (Rüger, 1996). The Jacobi method referred to as "jac" simultaneously diagonalizes the shifted correlation matrices of the sphered data. This method can only find an orthogonal matrix and has to rely on correct sphering. A comparison between the two plots shows that the multi-shift algorithms are able to perform much better than even the optimal single-shift method. For low to medium noise levels this is even the case when using the standard sphering method combined with the gradient descent algorithm. The advantage of the noise robust sphering method, compared to the standard sphering, is obvious: the reconstruction error stays very low for all evaluated noise levels and for both "jac" and "dpa". The gradient descent technique is more robust against erroneous sphering than the Jacobi method.

Figure 7.16 shows results which were produced using a single mixing matrix, but further simulations show that the algorithms compare qualitatively similar when using mixing matrices with condition numbers between 2 and 10.

# Chapter 8

# REAL WORLD APPLICATIONS OF SOURCE SEPARATION TECHNIQUES

## 1. Nonlinear Model-Analysis of Calcium Imaging Signals from the Honeybee Brain

### 1.1 Background

During the last couple of years, optical imaging of neuronal population activity using $Ca^{2+}$-sensitive dyes (Tsien, 1980; Poenie, 1992; Galizia et al., 1997) has become an important experimental tool for the investigation of neuronal response properties and coding strategies. As mentioned in chapter 4, section 1.2, $Ca^{2+}$ imaging uses the fact that postsynaptic intracellular $Ca^{2+}$ concentration often increases when the cells are activated, and that a presynaptic calcium-influx is necessary for transmitter release. Neural population activity is recorded by introducing a $Ca^{2+}$-sensitive fluorescent dye to the intracellular neuronal space of the considered tissue, illuminating it, and recording the amount of fluorescent light over space and time during neuronal stimulation. The resulting change in the two-dimensional pattern of fluorescent light intensity is taken as a measure of the pattern of neuronal population activity. Recently, this technique has been successfully applied to characterize important aspects of neuronal olfactory coding in the antennal lobe of honeybees (Joerges et al., 1997; Galizia et al., 1998; Galizia et al., 1999; Sachse et al., 1999).

The optical signal obtained in $Ca^{2+}$ imaging experiments represents a composite signal (cf. also figure 8.2), the structure and the components of which have not yet been analyzed in depth. First of all, the signal contains at least one component that is due to photobleaching of the dye. Second, the intracellular $Ca^{2+}$-concentration generally can change due to several mechanisms, not all of which are necessarily linked to neuronal activity. Third, autofluorescence may form another component of the fluorescence signal. This implies that the change of the intracellular $Ca^{2+}$-concentration, which gives rise to the neuronal activity estimate, in general cannot be determined simply from the amplitude of the measured fluorescence composite signal, because this signal changes due to several causes. We face the issues of $(i)$ the identification of the different signal components from the measured mixture, and $(ii)$ their separation and quantification.

In this section we demonstrate that a regression-based analysis of $Ca^{2+}$-imaging data by nonlinear parametric models (cf. chapter 6, section 2 provides a powerful tool for the identification and mapping of $Ca^{2+}$-signal components (Stetter et al., 2000b). We fit both individual time-series and complete image stacks obtained from $Ca^{2+}$-imaging experiments of the antennal lobe of honeybees (*Apis mellifera*) by NPMs and obtain the following key-results: $(i)$ Besides photobleaching, the time series averaged over active regions of the antennal lobe contain both a fast and a slow component. $(ii)$ For subsequent trials during an experiment, the amplitude of the fast component decreases over time, whereas the amplitude of the slow component increases. $(iii)$ The spatial distributions of the amplitudes of both components in response to stimulation are not identical but differ from each other. These findings provide for the first time a consistent interpretation of the fluorescent signals, yet arise the important issue of which neuronal processes underlie the two signal components we have observed.

## 1.2 Animal Preparation, Staining, and Data Collection

In-vivo calcium recordings have been performed as described in (Galizia et al., 1998). After capture from the hive, bees were quickly chilled for anaesthetisation and fixed in a Plexiglas chamber. The head capsule was opened and the brain was floated in a solution of Calcium Green 1 or 2 AM (Molecular Probes, Eugene; 50 $\mu$g dye were first dissolved in 50 $\mu$l Pluronic in DMSO and then diluted in 950 $\mu$l Ringer, 130 mM NaCl, 6 mM KCl, 4 mM $MgCl_2$, 5 mM $CaCl_2$, 160 mM sucrose, 25 mM glucose, 10 mM HEPES, pH 6.7, 500 mOsmol). After 1 h staining, the brain was rinsed in fresh Ringer, and the recording chamber placed under the microscope with constant Ringer perfusion (1 ml/min) at room temperature (22 deg Celsius). At all stages, great care was taken not to wet the antennae.

Images of a $240 \times 250\mu$m square region of the antennal lobe were taken with a 12 bit CCD camera (Photometrics CH250A, Tucson, Arizona) at a rate of 2 images per second, with 240 ms exposure time per image. For every trial, either $M = 40$ or $M = 50$ images, corresponding to 20 or 25 seconds, were recorded, while an odor was applied to the antennae from $t = 3$ s until $t = 5$ s after the beginning of the image collection. Before storage, each image was spatially binned to a final size of $P_x \times P_y = 49 \times 51$ pixels. After each trial, a 35 or 40 second recovery period without illumination preceded the next trial. One single experiment consisted of six trials with odor followed by one trial

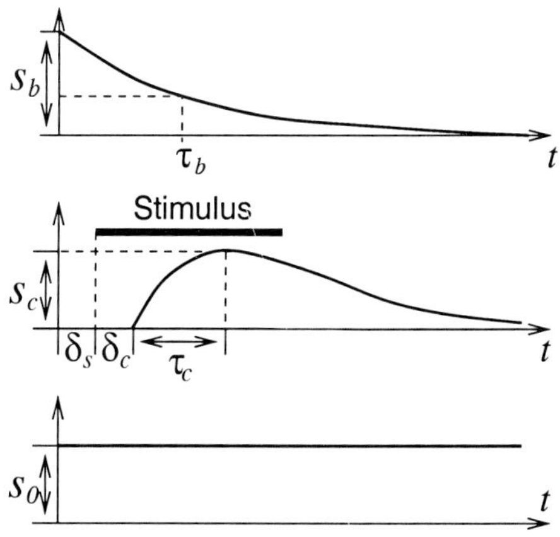


*Figure 8.1.* Families of model functions used for the NPM. Top: Exponential function for photobleaching components, Center: $\alpha$-function for stimulus-evoked components, bottom: background component of fluorescence. $s_b, s_c, s_0$ are linear amplitude parameters, $\tau_b, \delta_c, \tau_c$ are nonlinear curve shape parameters.

with plain air as stimulus (air trial). Usually, many experiments were carried out for a given animal.

The data format of an image stack as collected during one individual trial corresponds to the format illustrated in figure 6.4.

## 1.3 Function Families for the NPM

For application to $Ca^{2+}$ imaging data, families of model functions $a_l(t; \Phi)$ for the NPM described in chapter 6, section 2 were chosen as follows (figure 8.1). Firstly, the fluorescence signal is subject to photobleaching, which might be caused by one or several different mechanisms. Photobleaching is described by an exponential function,

$$a_b(t; \tau_{b,i}) = \exp\left(-\frac{t}{\tau_{b,i}}\right) \tag{8.1}$$

(figure 8.1, top), where $i$ indexes different photobleaching mechanisms and $\tau_{b,i}$ are the corresponding time constants. Stimulus-evoked changes in fluorescence contain both a $Ca^{2+}$ release mechanism from internal stores or the influx of extracellular calcium (Tsien and Tsien, 1990) and a subsequent buffer mechanism (Friel and Tsien, 1992). When neurons become active during a small period of time, a first-order calcium release mechanism will increase the concentration $c(t)$ of free $Ca^{2+}$ according to $c(t) = c_0 + (c_\infty - c_0)(1 - \exp(-t/\tau))$ (with some time constant $\tau$), which can be well approximated by a linear function during the early phase. At the same time, a first-order buffer mechanism tries to restore the resting level by an exponential decay. Hence, if we assume first order reaction kinetics to hold for the $Ca^{2+}$-release and buffer mechanisms, the alpha-function,

$$a_c(t; \delta_{c,i}, \tau_{c,i}) = a_0 \frac{t - \delta_{c,i} - \delta_s}{\tau_{c,i}} \exp\left(-\frac{t - \delta_{c,i} - \delta_s}{\tau_{c,i}}\right) \quad (8.2)$$

(figure 8.1b, center) with its initial linear increase and its late exponential decay represents a good phenomenological description of a stimulus-evoked component. In eq. (8.2), $\delta_s$ denotes the time of the stimulus onset, $\delta_{c,i}$ is the response latency and $\tau_{c,i}$ denotes the rise-time of the response for the $i$-th stimulus-related component. $a_0$ is a constant which normalizes the function to a maximum value of 1. Finally, a constant function describes the background fluorescence (figure 8.1b, bottom). For example, a nonlinear model with one bleaching component and two stimulus-related components is described by the parameter sets $\mathbf{s} = (s_0, s_b, s_{c1}, s_{c2})$, where $s_0$ is the amplitude of the background fluorescence, and $\Phi = (\tau_b, \delta_{c1}, \tau_{c1}, \delta_{c2}, \tau_{c2})$.

For the model functions and parameterizations used, the cost function turned out to be relatively smooth and the optimization by standard numerical gradient descent quickly converged to a unique result. This indicated, that the optimization process was not disturbed by local minima. Initial values of nonlinear parameters were set as estimated by visual inspection from the $\Delta F/F$ image (figure 8.2) to $\delta_{c1} = 0.25$ s, $\delta_{c2} = 2$ s, $\tau_{c1}^{-1} = 0.10$ s $\tau_{c2}^{-1} = 0.05$ s. The gradient was calculated numerically using step sizes of $(0.01; 0.01; 0.0004; 0.0002)$ for the four parameters. The step size for the parameter update was adapted as follows: If the gradient along a direction changed sign in two subsequent steps, the step size was decreased by a factor of two, if its sign was preserved for three subsequent update steps, the step size was multiplied by a factor of 1.25. Programs were implemented in Matlab on a SUN ULTRA 10 workstation. Computation time was approximately 1 sec per pixel time series.

## 1.4 Identification of Signal Components from Mean Time Series

For the identification of different components of fluorescence, nonlinear model fits were first applied to mean time series taken over an active region, i.e. over a region where the fluorescence increased in response to an odor. The approximate extensions of active regions were determined prior to the analysis from the $\Delta F/F$ image (Joerges et al., 1997), which represents a simple estimator of the relative change in fluorescence due to olfactory stimulation: It was obtained as follows:

1 Use a stimulus-trial $x_m(\mathbf{r})$.

2 Calculate the mean fluorescence before stimulation: $F(\mathbf{r}) = \frac{1}{6}\sum_{m=1}^{6} x_m(\mathbf{r})$

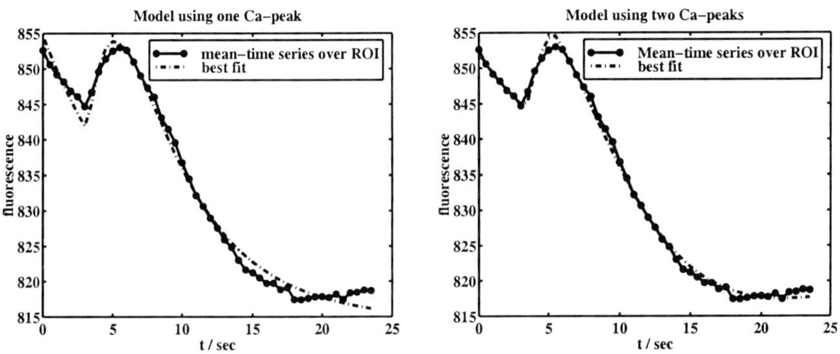


(a) (b)

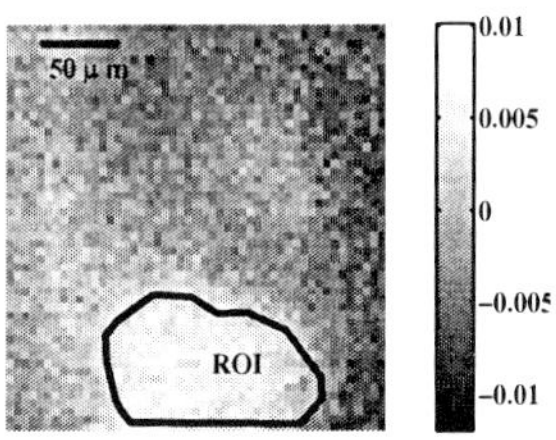


*Figure 8.2.* Nonlinear model analysis of a mean time series. **(a)** Mean time series (black circles) and the best nonlinear fit using one photobleaching and one stimulus-evoked component, and **(b)** best fit for one bleaching and two stimulus-evoked components. Left: $\Delta F/F$ image together with the region of interest over which the mean-time series was calculated.

3 Calculate the mean fluorescence during maximal response: $F_r(\mathbf{r}) = \frac{1}{4}\sum_{m=10}^{13} x_m(\mathbf{r})$

4 $\Delta F/F(\mathbf{r}) = (F_r(\mathbf{r}) - F(\mathbf{r}))/\mathbf{F}(\mathbf{r})$.

The gray scale image in figure 8.2 shows the $\Delta F/F$-image of an antennal lobe for the sum over 7 trials under stimulation with hexanol, showing approximately 1 % change in fluorescence in the central lower part of the image. This region corresponds to a set of glomeruli, (the functional units of the antennal lobe), which have been driven by the presentation of hexanol. $\Delta F/F$ images represent the common way of analyzing $Ca^{2+}$-imaging data. Here, however, we aim at a more detailed characterization of the fluorescence signal in space and time.

The solid lines with black circles in figures 8.2a and 8.2b represent the mean time series obtained by averaging over the pixel time series within the region of interest marked in the $\Delta F/F$ image in figure 8.2. The negative slope prior to stimulation indicates photobleaching and the increase starting from $t = 3$ s corresponds to a stimulated response. The dashed line in figure 8.2a shows the best fit, if one bleaching-component and one stimulus-evoked component are used. The photobleaching component is overestimated, as can be seen from its high slope in the early phase. This poor performance is caused by a negative undershoot in the late phase of the data. Figure 8.2b (dashed) shows the best fit for one bleaching and two different stimulus-evoked components. The second

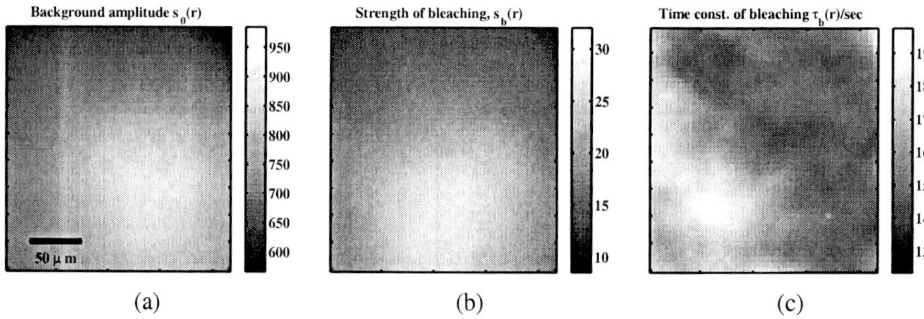


*Figure 8.7.* Statistical parametric maps from an air trial. **(a)** The background fluorescence $s_0(\mathbf{r})$ **(b)** The strength of the photobleaching component $s_b(\mathbf{r})$ and **(c)** the time-constant $\tau_b(\mathbf{r})$ of the photobleaching.

lus evoked and cannot be explained as autofluorescence signals. The fast and slow components either are related to the same $Ca^{2+}$-dependent mechanisms or are causally related to each other, because pharmacological treatment always affects both signals.

## 1.5 Analysis of Spatial Distributions

Proceeding from the identified signal components, we can apply nonlinear models pixelwise to all pixel time series of complete image stacks. Because individual pixel time series were more noisy than mean time series, results turned out to improve, if we first corrected for photobleaching and afterwards fitted a model using only the background component and the two signal components. For the correction, we first fitted a NPM with one background and one photobleaching component to the air trial of each experiment. Figure 8.7 shows the resulting statistical parameter maps for the background fluorescence $s_0(\mathbf{r})$ and the strength $s_b(\mathbf{r})$ together with the time constant $\tau_b(\mathbf{r})$ of the photobleaching component. From these experiments we learn that bleaching is well described by a single exponential decay. Typically, its strength varies over space by $(\Delta s_b/s_{b,\text{mean}}) \approx \pm 0.5$, and its time-constant by $(\Delta \tau_b/\tau_{b,\text{mean}}) \approx \pm 0.25$, where the index $mean$ marks the mean value of a quantity over all pixels of the image. Once the best model for the air trial was fitted, the model instead of the air trial was used to correct for photobleaching in the stimulated trial by subtraction. This procedure avoids adding noise from the air trial to the image stack to be analyzed. After correction, each pixel time series of a trial was fitted using a GLM (cf. chapter 6, section 1) with a background component and the model functions of the fast and the slow components with fixed nonlinear parameters as determined from the corresponding mean time series. Because the nonlinear curve shape parameters were kept fixed for each image stack, only three amplitude parameters had to be fitted for each time series.

Figure 8.8 shows the best fit as described above for the same data set as used in figure 8.2. The left column contains every forth frame of the bleach corrected original image stack, the middle column the best model fit and the right column shows the residual, which is the difference between the data and the model. Firstly, the residual contains little structure, which indicates that all the relevant structure in the data is successfully captured by the model. Because the curve shape of the model functions is kept constant, this means that the whole spatiotemporal signal can be explained by the presence of only two signal components. In other words, the temporal dynamics of the signal components are not only constant between odors and animals, but are also constant over space within the same animal. Secondly, it becomes visible that the spatial patterns of the fast and slow components are different from each other. This property is demonstrated more strikingly in figures 8.9 and 8.10.

In figure 8.9, the statistical parametric maps 8.9a and 8.9b and the corresponding Z-scores 8.9d and 8.9e are shown for the fit of figure 8.8. The signals extracted by the NPM are much less noisy than the signals extracted by $\Delta F/F$ analysis, and by the Z-score we can quantify our confidence in them. Note that these maps result only from a temporal analysis, i.e. no spatial processing such as smoothing has been applied. The image figure 8.9f is the difference between the strengths of the fast and slow components. It is nonzero, in the case shown amounts to approximately 30 % of the signal components itself and is spatially modulated. We conclude that we have similar spatial distributions but different details in structure in the fast and slow components.

the spatial distributions of the fast and the slow component are similar, but can differ from each other in the details of their structure.

This effect is demonstrated again in figure 8.10, which shows the spatial distributions of the fast and slow responses for a different animal, and for three different stimulus conditions. In all cases, the fast and the slow component for a given odor are more similar to each other than to any component of another odor, but they are not identical. This is demonstrated by the difference maps between the fast and the slow components in the right column of figure 8.10. The fast and slow component usually show a similar pattern of active glomeruli, however the intensity distributions of both components are different. Sometimes, this difference is large for an individual glomerulus, which then appears only in one of the components (e.g. the activity spot at the right edge in the top row images). The second aspect shown in figure 8.10 is the approximate linearity of the fluorescence signals. From top to bottom, the first three rows show responses to limonene, hexanol, and a symmetric mixture of both odors with the same concentrations. The three maps directly below the horizontal line are obtained by a linear superposition of the maps for stimulation with the pure odors. If

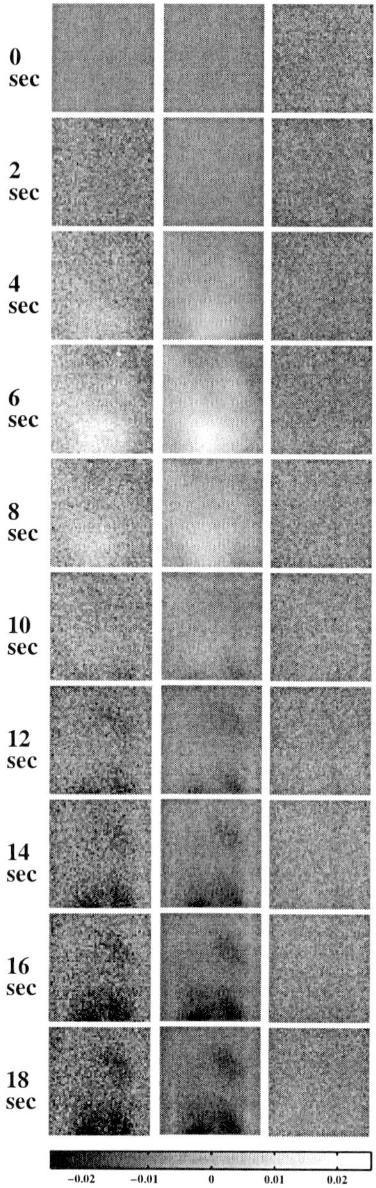


*Figure 8.8.* Spatiotemporal fit of a whole image stack. **(Left)**: Same image stack as analyzed in figure 8.2 after model-based correction for photobleaching (every forth frame shown).**(Middle)**: Best pixelwise fit of the data using the two signal components as determined from mean-time series. **(Right)**: Residual image stack (model subtracted from data). The residual contains little structure indicating that the model successfully extracts all the relevant information from the data.

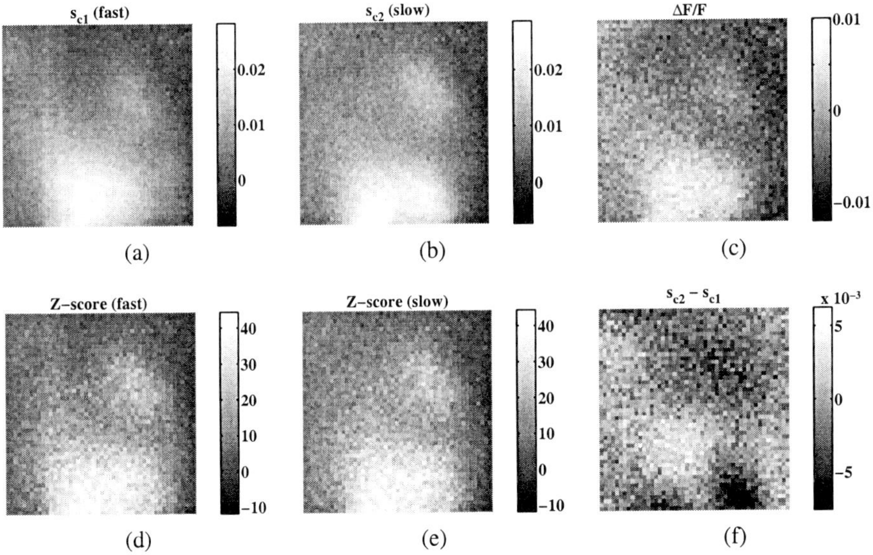


*Figure 8.9.* Statistical parametric maps of the fast **(a)** and slow **(b)** components for the fit shown in figure 8.8. Both maps are much less noisy than the corresponding $\Delta F/F$ image **(c)**. **(d)** and **(e)** Z-score distributions of the fast and slow components in (a) and (b), respectively. Some glomerular regions are active with high significance. **(f)** The difference between the fast and slow component amounts to approximately 30 % of each component. Its spatial modulation indicates, that the fast and slow components are differently distributed over space.

the fluorescence signal behaved linearly, the mixture of odors would evoke the same response than the superposition of the responses to the pure odors. A comparison of the third and fourth row from top with each other demonstrates an expressed similarity of the spatial patterns for both cases, which corroborates the linearity assumption. The total strength is somewhat weaker in the response to mixed odors than expected by a linear behavior, which hints towards the presence of lateral inhibition in the antennal lobe. This difference in strength is demonstrated by the negative regions in the two two bottom plots, which display the difference between the response to odor mix and the superposition.

## 1.6 Discussion and Conclusions

Sets of nonlinear model functions were fitted to the time courses of fluorescence signals from $Ca^{2+}$-imaging experiments. Using this technique, we were able to identify two well-defined and well-distinguishable dynamical components of the stimulus-related fluorescence signal – one signal corresponds to a relatively fast increase in local fluorescence, whereas the second components reflects a slower decrease below background fluorescence. The temporal dynamics of both components are very well constant over space as well as over

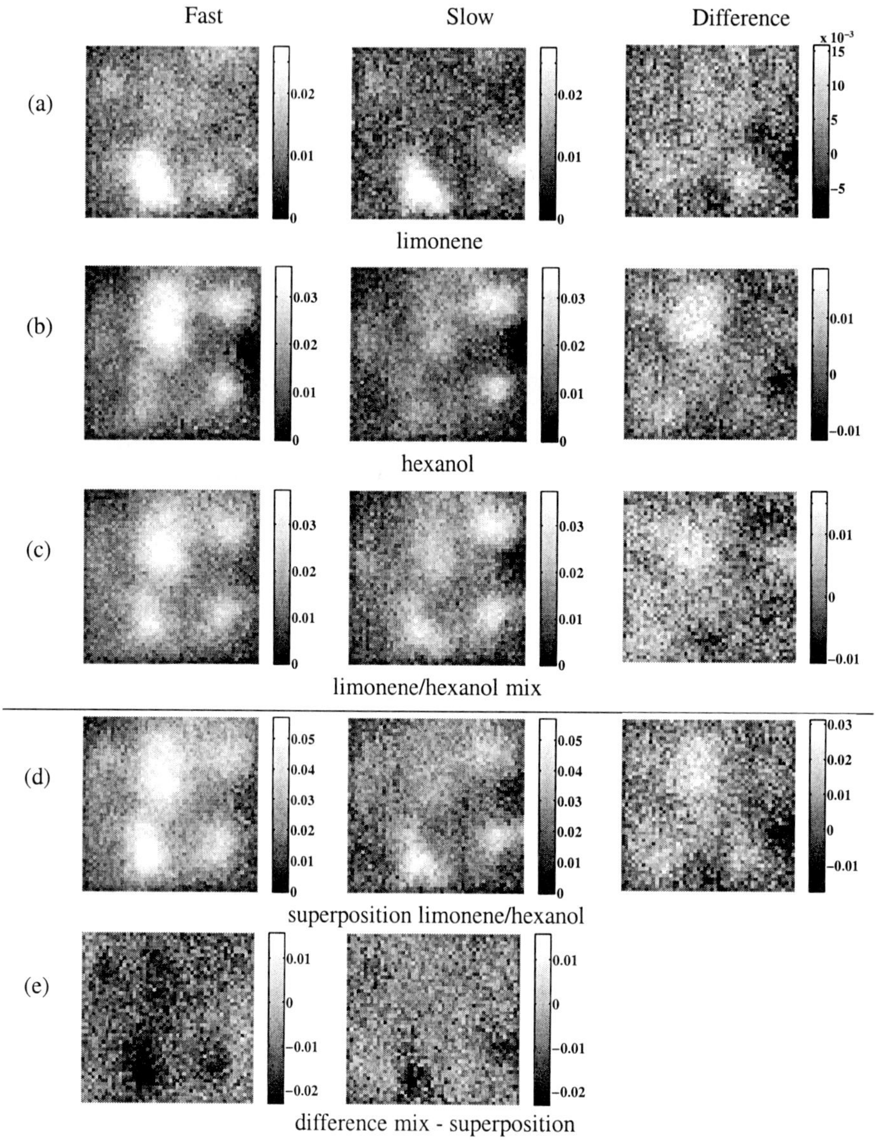


*Figure 8.10.* Different examples for the spatial maps of fast **(left)** and slow **(middle)** components and their differences **(right)**. The respective patterns for stimulation with limonene **(a)**, hexanol **(b)** and a mixture of both odors **(c)** are similar but not identical for the fast and slow components. **(d)** Linear superposition of responses to pure odors and **(e)** the difference between the response to the mix and the superposition. Both mix and superposition patterns are similar to each other in space, however the response to the mix is slightly smaller than expected by a linear superposition.

different trials, odors, and animals, but the spatial amplitude distributions of both components can differ considerably from each other.

The goal of $Ca^{2+}$ imaging in this system is to infer the spatial or spatiotemporal patterns of neuronal population activity in response to olfactory stimulation exploiting the relationship between neuronal activity and the amount of intracellular $Ca^{2+}$. This $Ca^{2+}$-level is monitored by measuring the change in fluorescence caused by a $Ca^{2+}$-sensitive dye. Our identification of two different components of stimulus-related fluorescence with slightly different spatial distributions rises the important issue of what is the relationship between each of the components and neuronal activity. Possibly the different components reflect different cellular mechanisms of $Ca^{2+}$ supply, which in turn might be related to different stages of olfactory coding. For example, the fast and slow components may reflect successive stages of olfactory processing in the antennal lobe. In this case, the differences of the spatial pattern of the two components would indicate that odors elicit not just a fixed set of activated glomeruli, but rather a slow sequence, with some glomeruli being activated with a delay. More biophysical work has to be done in order to specify the origin of each component as well as their relationship with neural activity. We hope that our characterization of the behavior of the responses over time and space can help addressing these questions.

Nevertheless, some speculations about the origin of both signals should be discussed. We observed that both signal components are reversibly and simultaneously removed $(i)$ by application of TTX and $(ii)$ by removal of free extracellular $Ca^{2+}$, both of which block neural activity in the antennal lobe (figure 8.6). $(iii)$ Also, both components are statistically insignificant in most air trials (in some air trials, weak stimulus-related components were observed, but those can be assigned to residual odor impurities in the air used and are not reproducible). The first and the third observation indicate that the fluorescence signals are coupled to neuronal activity, whereas the second test hints towards a coupling either on activity or on a $Ca^{2+}$ influx. Alternatively, one might argue that autofluorescence contributes to the measured change in fluorescence and could provide one of the signal components. This interpretation is questionable, because in experiments carried out without staining, no significant fluorescence signals have been observed. It can be concluded that neuronal activity and the $Ca^{2+}$-sensitive fluorescent dye are necessary for the generation of either component.

Based on these observations it is still unclear what is the origin of the components and why they behave differently over space. For example, the slow signal component always had a negative amplitude, which apparently reflects a strong decrease in intracellular $Ca^{2+}$ concentration. However, $Ca^{2+}$ ions are very effectively buffered in the intracellular space, resulting in a very low concentration in the range of 100 nM (Bygrave and Benedetti, 1996). Therefore, even a complete removal of free $Ca^{2+}$ could not account for the strong undershoot observed. One alternative possibility of its origin could be a compartmentalization of the fluorescent dye into the endoplasmatic reticulum (ER), which operates as a $Ca^{2+}$-reservoir. This interpretation is corroborated by the increase of the slow component over time, which could reflect the increase of compartmentalized dye. Pharmacological studies have to be undertaken to prove this hypothesis, according to which the second component would reflect a $Ca^{2+}$ release out of the ER (Bygrave and Benedetti, 1996). Our nonlinear model would then allow to segregate the contribution of extracellular calcium influx (first component) and intracellular calcium release (slow component) in a single measurement. Alternatively to this hypothesis, the two signals could be generated by different mechanisms in different cell types (Verkhratsky and Kettenmann, 1996).

The present method characterizes fluorescence signals with much higher accuracy than $\Delta F/F$ images, and additionally provides confidence levels. By use of these properties it becomes possible to address new aspects of neuronal encoding of olfactory stimuli. For example, the strength and the type of nonlinearities in the code of composite stimuli (Joerges et al., 1997) can be quantified and tested for significance. Also, it becomes possible to identify the positions of individual glomeruli directly from sets of $Ca^{2+}$ imaging experiments by use of the Z-score. Based on those methods, it will become possible to quantitatively characterize the vector coding strategy, by which olfactory stimuli are represented in the antennal lobe of honeybees.

## 2. Principal Component Analysis for Optical Imaging of Intrinsic Signals

In this section we demonstrate, that PCA already represents a powerful tool for partial or even complete source separation of frame stacks from optical imaging of intrinsic signals, provided the method is correctly applied. The data analyzed will include data sets which have been shown already in chapter 5. There, we have seen that these data sets (e.g. figure 5.6) cannot be satisfactorily analyzed by established heuristic methods. Here we will observe that PCA often outperforms differential imaging or bandpass filtering in its source separation abilities, and will characterize to what extent and under which conditions this is the case.

### 2.1 Animal Preparation and Data Collection

Optical imaging of intrinsic signals was performed on area 17 of ferrets and cats. Techniques for surgery and animal preparation have been described in detail in (Bonhoeffer and Grinvald, 1996) and (Sengpiel et al., 1998b). For optical imaging, the cortex was illuminated using bandpass-filtered light of $605 \pm 10$

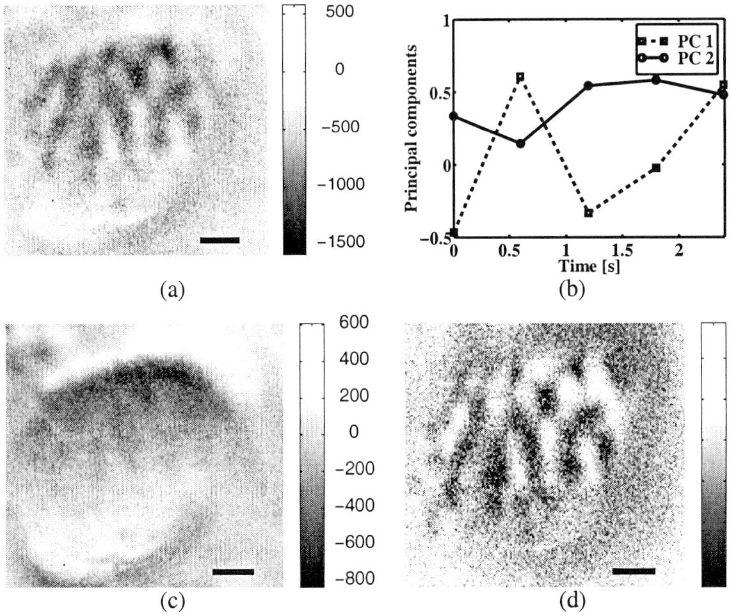


*Figure 8.11.* Removal of a time-dependent artifact using PCA. **(a)** Summed differential image $(0-90$ deg orientations without first-frame analysis) recorded from ferret visual cortex. **(b)** The time series of the first (PC 1) and second (PC 2) principal components of the differential image stack. Eigenvalues (variances) were: $\lambda_1 = 45157$, $\lambda_2 = 7489$. **(c)** First and **(d)** second principal image obtained from the two time series in (b) using eq. (7.8). Scale bar: 1 mm; the gray scales denote gray values of pixels. Its ranges were determined as the mean $\pm$ three times the standard deviation for each image.

nm. Images of intrinsic signals in response to visual stimulation were recorded using a cooled slow-scan CCD-camera (ORA2001, Optical Imaging, Germantown, NY). Stimuli consisted of moving oriented bar gratings. If not specified explicitly, for each trial five frames of 600 ms duration were recorded during each 3 s stimulus presentation, followed by an inter-stimulus interval of 7 seconds, in which the next stimulus was presented stationary. Single-condition and difference image stacks $\mathbf{X}$ with $(\mathbf{X})_{m,\mathbf{r}} = x_m(\mathbf{r})$ were obtained by pixel-wise summation over 16 trials eventually followed by first frame analysis, as described in chapter 5.

## 2.2 PCA on a Differential Image Stack

Figure 8.11 demonstrates the removal of a time-dependent artifact from a differential image stack using PCA. The data were recorded optically from ferret visual cortex using 0 deg and 90 deg oriented stimuli and represent the same image stack as shown in figure 5.6. Figure 8.11a shows the summed differential

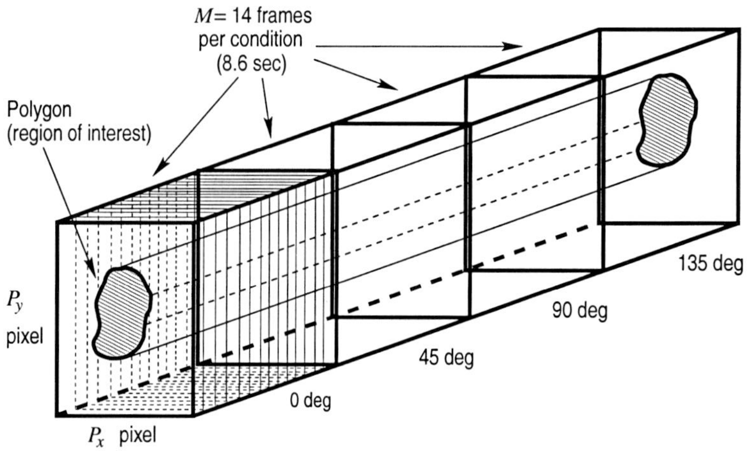


*Figure 8.12.* Combination of image stacks for improved performance of PCA. $K$ different but not necessarily orthogonal single-condition stacks are combined to yield a long image sequence with $K\ M$ pixels per time series. The numbers shown ($K = 4$, $M = 14$) refer to the particular data set analyzed in figures 8.13 and 8.14.

image (without first frame analysis), which contains both the mapping signal and a large, roughly annular artifact. The annular patterns resulted from a time-dependent artifact which cannot be removed by differential imaging (cf. section 2.1). In figure 8.11b, the first two principal components, $\mathbf{v}_1$ and $\mathbf{v}_2$, are plotted as time-series. The first principal time series has a large variance and a complicated time course, which describes the time-dependence of the artifact, while the second principal component has a smaller variance and a time course which can be interpreted as the behavior of the mapping signal. The corresponding principal images $y_1(\mathbf{r})$ and $y_2(\mathbf{r})$ are given in figures 8.11c and 8.11d. They demonstrate that the mapping signal (figure 8.11d) can be almost completely separated from the time-dependent artifact (figure 8.11c) by PCA and corroborate the previous interpretation of the time series. Note, that the separation by PCA was possible, because the artifact and the mapping signal differed strongly in their variances over space and had approximately orthogonal time series. Figure 8.11c still contains a small fraction of the mapping signal, which is probably due to the presence of white noise and to small deviations of the real time series from orthogonality.

## 2.3 PCA on Sequences of Combined Image Stacks

The time series of the global and the mapping signals are both locked to the stimulus. As a consequence they are probably not orthogonal to each other and are not separable by PCA. For the separation of the global and the mapping signal using PCA, we generate a combined data stack, which contains $K > 1$ single-condition image stacks for different stimuli in a sequence such as to form a long time series of $KM$ images (figure 8.12). In this combined stack, the global signal shows a similar response irrespective of the stimulus type, whereas the time-course of the mapping signal depends on the stimulus. Therefore, the time series of both signals are now likely to be nearly orthogonal. Additionally, global signals are known to be stronger than the mapping signal components, their variance can be expected to be higher. We can summarize that in a combined image stack, the time series for global and mapping signal sources are likely to be orthogonal and to show different variance over the ensemble of pixels. Because of these two properties, however, the source separation problem becomes treatable by PCA.

Figure 8.13 shows the first nine principal time series (figure 8.13a), the corresponding principal images (figure 8.13b), and the Scree plot (the logarithm of the principal component eigenvalues ordered in size) for a combined data stack, which had been generated by stacking four single-condition image sets ($M = 14$ images each, 8 s stimulus duration, 2 s time between stimuli) for stimuli with orientations 0, 45, 90, and 135 deg and performing first frame analysis (figure 8.13). The Scree-plot shows that only the first 7-8 principal components carry signals, whereas the higher components, which show roughly uniform variances, carry noise. This agrees with the noisy appearance of the 9th principal time series and images. The time-dependences of the first three principal components from top are nearly identical for each stimulus presentation irrespective of the stimulus type. Using their spatial patterns they can be assigned to the global signal (top image) and to signals from larger blood vessels vascular patterns and illumination artifacts (second and third image). The fourth to seventh principal components contain linear combinations of the mapping signals for the different stimulus-conditions, which appear as periodic patterns in the cranial window region of the corresponding images. Note, that PCA cannot separate different single-condition responses from each other, because their variances usually are similar. It is not useful to assign any biological meaning to the individual principal components that capture the mapping signal, but they can be used to reconstruct a cleaned data set with reduced noise and global signal components.

Signal extraction and noise reduction of the data set in figure 8.13 can be achieved by reconstructing the data set, cf. eq. (7.9), using only the principal components 4-7. Discarding the higher principal components reduces noise, while discarding the first three principal components removes part of the global signal and the blood vessel artifacts. Single condition images can be obtained by pixel-wise summation of all resulting images for a given stimulus condition or by fitting a linear model to the cleaned time series. Figure 8.14a shows the four single condition images of the data set of figure 8.13 before PCA, where

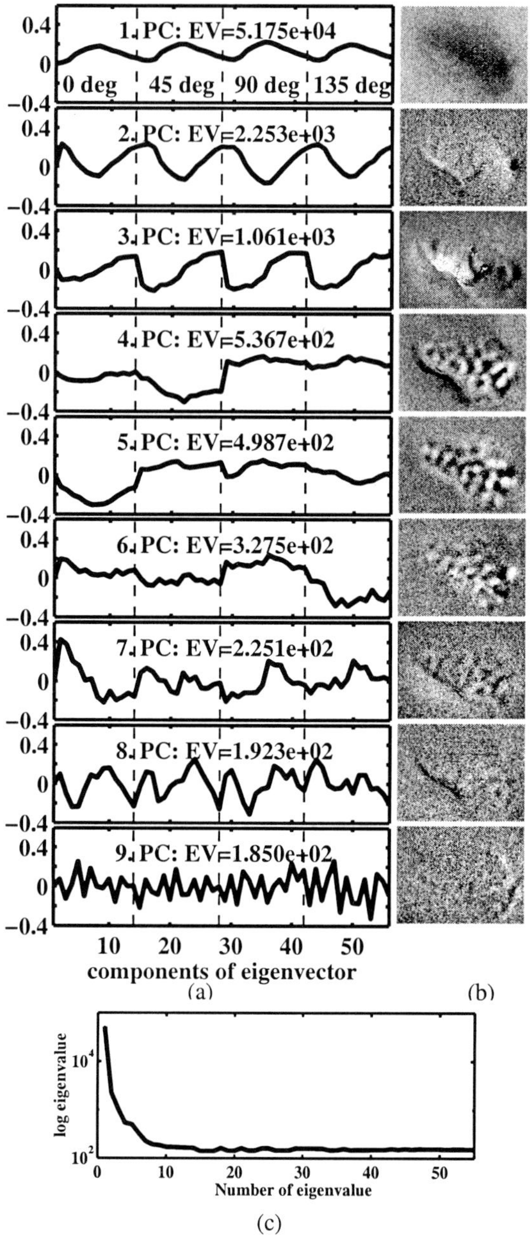


*Figure 8.13.* **(a)** The first nine principal time series and **(b)** the corresponding principal images, normalized to unit variance, for an image stack from cat area 17 (60d) after combination of four single condition stacks (0,45,90,135 deg) and first-frame analysis (gray scale: $\pm 3$). Legends provide the eigenvalues (variances). The first three components capture the global signal and blood vessel patterns, whereas the mapping signal is concentrated in components 4-7. **(c)** Logarithm of the eigenvalues sorted by size (Scree plot). All but the first 7-8 principal components contain noise.

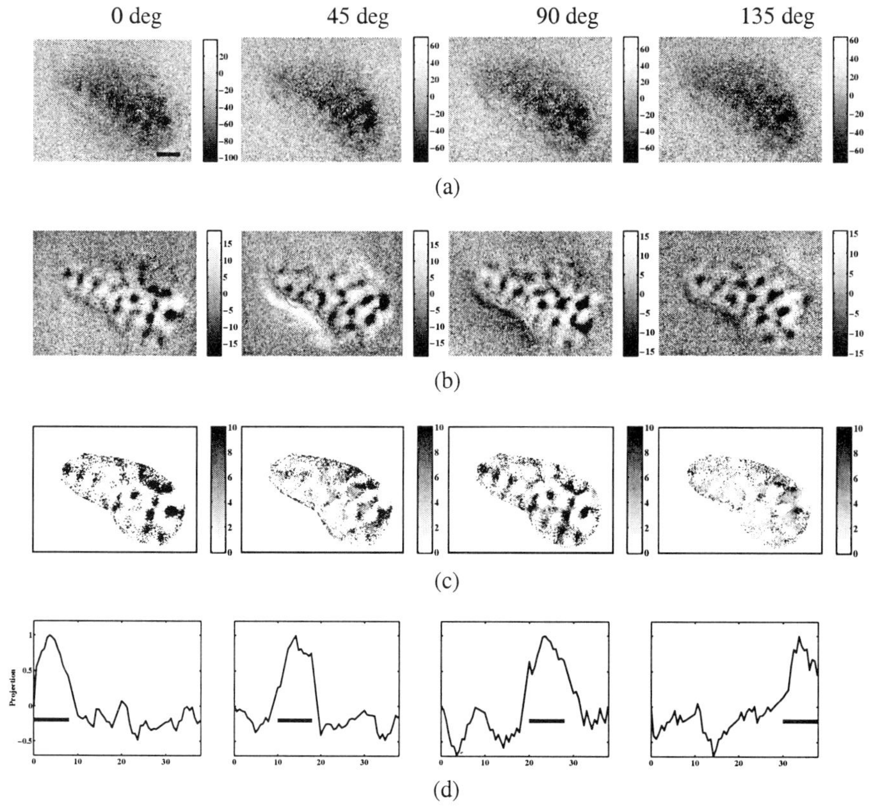


*Figure 8.14.* Reconstruction of the mapping signal for the data stack of figure 8.13. **(a)** Single-condition images (from left to right: 0 deg, 45 deg, 90 deg, 135 deg) after first-frame analysis. **(b)** Single-condition images for the cleaned data set using principal components 4-7 of figure 8.13. Scale bars: 1 mm. **(c)** Z-score for the data in (b). Regions of high significance of the mapping signal are shown in black. **(d)** Projection of the reconstructed images (b) onto the raw data. Bars mark stimulus duration. Color bar ranges were determined as the mean ± three times the standard deviation for all except Z-score images.

both the global signal (diffuse background) and the mapping signal (small dark blobs) are mixed together. Figure 8.14b displays the reconstruction of the same single-condition image after PCA and noise removal. The reconstruction has been done by determining the best curve-shape of an alpha function from a mean time series of the cleaned stack using a NPM, followed by a pixelwise application of a linear model using the best alpha function. The statistical parametric map of the GLM provides the reconstructed single condition image and the Z-score its significance areas. In the reconstructions, figure 8.14b, the global signal is drastically reduced compared to the original data, whereas the mapping signal is preserved and concentrated.

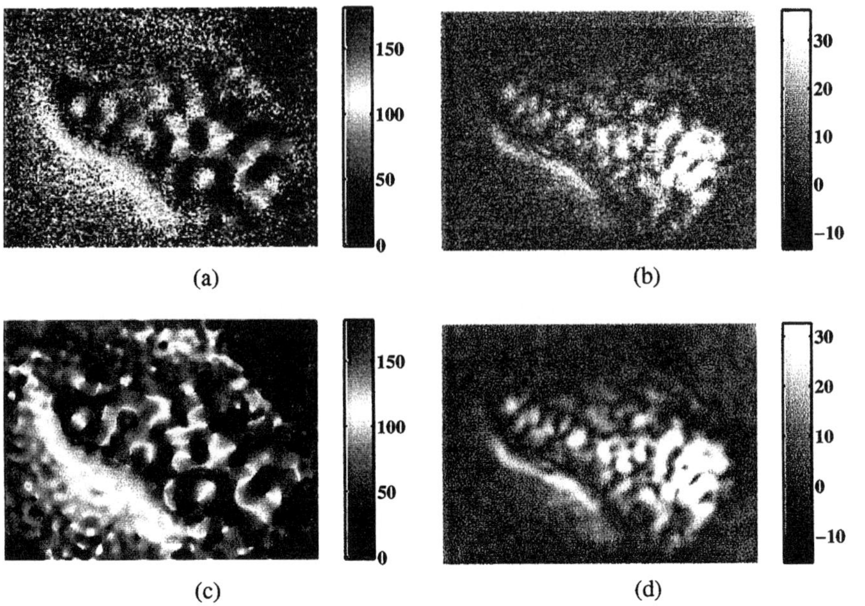


*Figure 8.15.* **(a)** Orientation (OR) map and **(b)** Orientation selectivity (OS) map obtained from the reconstructed single-condition maps of figure 8.14 without postprocessing. **(c)** and **(d)** same maps after lowpass-filtering (cutoff wavelength $\lambda = 0.27$ mm. Colors mark deg of preferred stimulus orientation. (*For color figure see Color Section,* p. xv)

Figure 8.14c displays the Z-score as a function of space within the region of the cranial window for the four reconstructed single condition maps. The dark islands in the Z-score mark the regions where the mapping signal differs significantly from zero. Figure 8.14d, finally, plots the time series $\mathbf{z}$ (normalized to unity), which results from the back-projection of the pattern $\mathbf{y}$ in figure 8.14b onto the original combined data stack: $\mathbf{z} = \mathbf{X}\ \mathbf{y}^T$. It measures, how much of the reconstructed image is contained in each individual frame of the data. Each estimated single-condition map appears most strongly within the part of the combined data stack, in which the corresponding stimulus condition has been presented (black bar). Also, the respective orthogonal stimulus-condition shows an anticorrelation, yet much weaker than the correlation in the correct stimulus-condition, which demonstrates that responses to orthogonal stimulus-conditions do not form completely disjunct patterns.

We can now combine the reconstructed single-condition stacks to obtain orientation maps for this animal. Following (Blasdel and Salama, 1986; Blasdel, 1992a; Stetter et al., 1993), a vector which describes the orientation selectivity of neurons at a pixel $\mathbf{r}$ can be obtained as follows: If we have $K$ single condition maps $s^k(\mathbf{r})$, which are obtained under stimulation with orientation $\theta_k$, $k =$

$1, ..., K$, the orientation preference vector $\mathbf{d}(\mathbf{r}) = (d_x, d_y)(\mathbf{r})$ can be obtained as

$$
\begin{aligned}
d_x(\mathbf{r}) &= \sum_{k=1}^{K} \cos(2\theta_k) s^k(\mathbf{r}) \\
d_y(\mathbf{r}) &= \sum_{k=1}^{K} \sin(2\theta_k) s^k(\mathbf{r}).
\end{aligned}
\tag{8.3}
$$

This formula corresponds to fitting a pi-periodic sinewave to the optical responses as functions of the stimulus orientation, $s^k(\mathbf{r}) \equiv s(\theta_k, \mathbf{r})$. The direction of the resulting vector represents twice the estimated preferred orientation and its length estimates the preference of one orientation over the orthogonal one:

$$\mathbf{d}(\mathbf{r}) = \mathrm{OS}(\mathbf{r}) \cdot (\cos(2\mathrm{OR}(\mathbf{r})), \sin(2\mathrm{OR}(\mathbf{r}))). \tag{8.4}$$

where $\mathrm{OR}(\mathbf{r})$ is defined as the orientation preference map, and $\mathrm{OS}(\mathbf{r})$ is the orientation selectivity map. For the special case of the four orientations used, the orientation preference vector becomes

$$\mathbf{d}(\mathbf{r}) = \left(s^{0\mathrm{deg}}(\mathbf{r}) - s^{90\mathrm{deg}}(\mathbf{r}), s^{45\mathrm{deg}}(\mathbf{r}) - s^{135\mathrm{deg}}(\mathbf{r})\right) \tag{8.5}$$

Figure 8.15a and 8.15b show the orientation-map (OR) and the orientation selectivity map (OS) that were obtained directly from the reconstructed single-condition images of figure 8.15. Both maps have been obtained without band-pass filtering and without differential imaging. The orientation map clearly shows the characteristic pinwheel patterns (cf. section2.4), but the difficulties and ambiguities related to heuristic methods for postprocessing of optical recordings (cf. sections 2.1 and 2.3) have been avoided. Figure 8.15c and 8.15d show the same maps after reduction of the high-frequency noise components with a lowpass filter (no highpass-filter has been applied). The resulting maps are now smoother, but a comparison with the unfiltered versions demonstrates that in contrast to bandpass filtering (figure 5.10), their structure has not changed. In addition, a mis-interpretation of filtered noise as an optical signal is unlikely, because the filter was chosen to yield a higher typical frequency of the filtered noise (image periphery) than of the optical signal (image center).

In summary, we showed that temporal PCA can be successfully used both for noise-reduction and for the removal of global signals and vessel artifacts. Because the global and the mapping signals show a different behavior over space and additionally differ in strength by approximately a factor of 5 (Bonhoeffer and Grinvald, 1996), their variances are strongly different in many cases and they can be separated using PCA. However, this requires application of PCA to a whole sequence of single-condition stacks. This mode of application extends recent work (Carmona et al., 1995; Cannestra et al., 1996; Everson et al., 1997;

Everson et al., 1998), where PCA has been applied on single condition or differential image stacks. These studies use PCA in order to reduce white noise and to extract prominent prototype time courses of the mixture of intrinsic signals, but different signal components are not separated.

# 3. Extended Spatial Decorrelation for Optical Imaging Data

PCA suffers from the drawbacks that it requires orthogonality of prototype time series, which is not necessarily given, and that ignores additional sources of knowledge besides the variances of time-series over the data set. In this section and following (Schießl et al., 1998; Schießl et al., 1999; Stetter et al., 2000c) we show that ESD analysis, which in contrast to PCA uses information about the source structure in space, improves the extraction of the mapping signal from the mixture and is also applicable to individual image stacks based on a single stimulus. The successful analysis of individual single-condition stack is of particular importance as future applications of functional imaging methods will include stimulus-conditions for which no orthogonal pairs exist. Again we will include single-condition data sets into our analysis which have been shown to be reluctant to established analysis methods (cf. chapter 5).

## 3.1 ESD on Differential Image Stacks

Figure 8.16 shows the result of an ESD analysis for an individual differential image stack. In figure 8.16a, we see a series of five images of a 0-90 deg differential image stack optically recorded from cat area 17 (time increases from top to bottom). It contains the oscillatory mapping signal that emerges superimposed to a global spatial pattern. Images were low-pass filtered with a cutoff frequency of (2.7 cycles/mm) in order to improve the performance of ESD (cf. (Schießl et al., 1999) or chapter 7), but no high-pass-filter was applied. Figure 8.16b lists the three leading principal images, and figure 8.16c the conjugate principal time series obtained from their back-projection. In this series, a movement artifact has been separated (third from bottom), but the mapping signal is still distributed across two images (first and second from bottom) and is still superimposed onto the global background. Figure 8.16d and 8.16e display three of the five spatial source patterns and their conjugate time series as estimated by ESD using the shift vector $\Delta \mathbf{r} = (5, 5)$ (the remaining two sources contain noise patterns). Now the mapping signal is completely concentrated in one source (second from bottom) and is separated from the global pattern (bottom). Figure 8.16f and 8.16g contain the conventional differential map and the mapping signal as provided by ESD analysis. A comparison reveals that ESD, in contrast to differential imaging, could remove the global signal and the movement artifact. In figure 8.16h, finally, the binarized absolute value of the

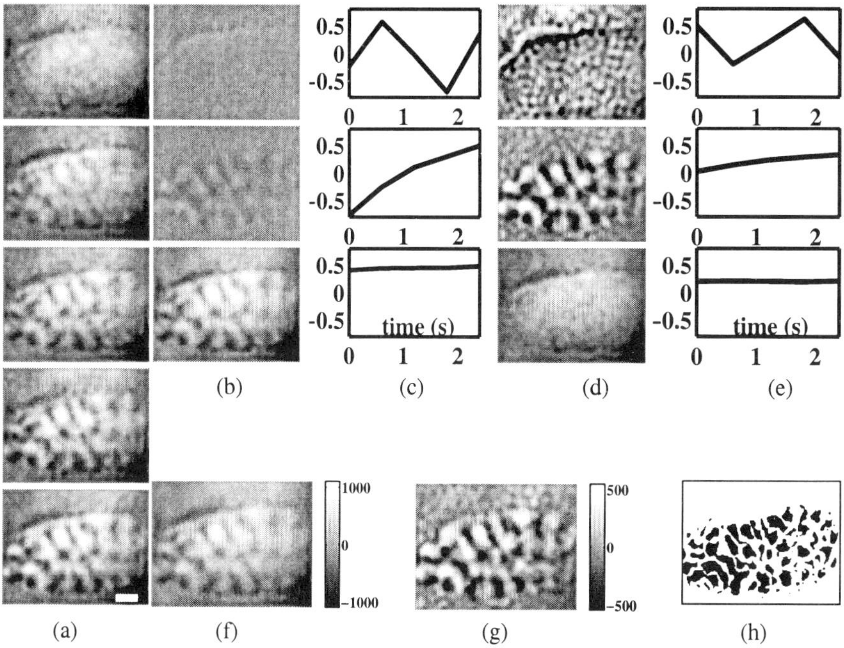


*Figure 8.16.* **(a)** Time series of images of a $0 - 90$ deg differential stack from cat area 17 (time interval: 600 ms). Images are low-pass filtered (cutoff: 2.7 $\text{mm}^{-1}$). Scale bar: 1 mm. Gray scale: $\pm 200$. **(b)** Leading three principal images (bottom = 1. PC; gray scale: $\pm 400$), and **(c)** conjugate time series for the data set shown in (a). Signal separation is incomplete. **(d)** Estimated spatial source patterns (arbitrary gray scale) and **(e)** conjugate time series after ESD analysis with $\Delta \mathbf{r} = (5, 5)$. The mapping signal is concentrated in the "center" source, and is separated from a global background and a movement artifact. **(f)** Resulting differential map obtained by summation of images in (a) **(g)** Resulting statistical parametric map obtained using a GLM. Model functions were the time series shown in the center and bottom graphs of (e). The global signal and the movement artifact have been successfully removed in this map. **(h)** Significance map for the statistical parametric map in (g). Black indicates regions for which the absolute Z-score was larger than 3.

Z-score is given for the map in figure 8.16g. Black marks regions, in which the absolute value of the Z-score for the middle time series is larger than three.

## 3.2 ESD of Single-Condition Stacks

ESD can also extract the mapping signal from an individual single condition stack. Figure 8.17a shows a 0 deg single-condition image for the same record as in figure 8.11. The data set consists of a mixture of the mapping signal, a global background and a time-dependent artifact. Figure 8.17b shows the best separation result for PCA (top), which is still very poor, and its conjugate time series (bottom). Figure 8.17c contains the best image (top) and time

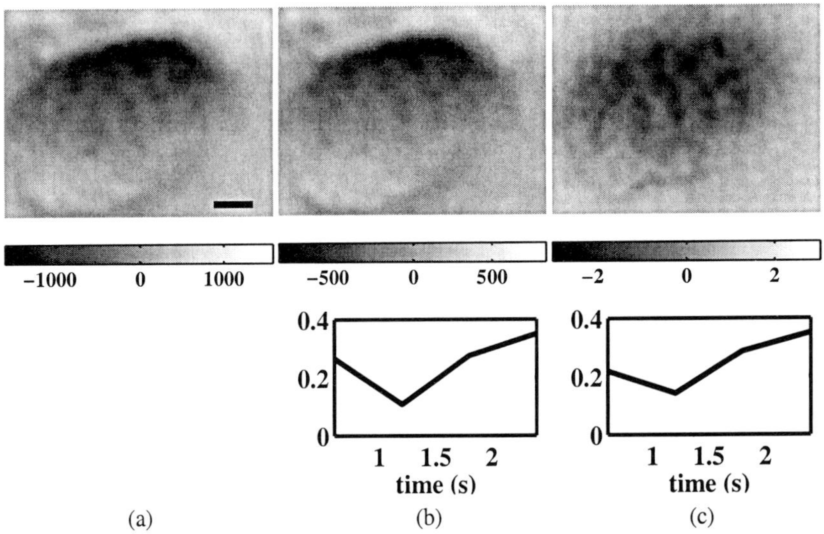


*Figure 8.17.* **(a)** 0 deg single condition maps of the same data set as in figure 8.11 and 5.6 after first frame analysis and pixelwise summation. **(b)** Best signal extraction using PCA. Top: principal image. Bottom: conjugate time series. **(c)** Spatial pattern obtained from ESD analysis (normalized to unit variance). Top: spatial pattern of the source. Bottom: conjugate time series. Images are low-pass filtered (cutoff: 2.7 $mm^{-1}$). Scale bar: 1 mm.

series (bottom) obtained from ESD. ESD successfully removes the artifact and achieves a strong yet not fully complete separation from the global signal.

## 3.3 Analysis of Highly Noisy Single-Condition Stacks

ESD can be used to separate even a weak mapping signal from other biological sources in the presence of noise. Figure 8.18a (top row) shows the single-condition stack of figure 5.3, from which the highest frequency components were removed by a lowpass filter with a cutoff-frequency of $\lambda = 0.094$ mm (40 cycles per image diameter). The goal for the analysis is to separate the mapping signal from the blood-vessel patterns and the global background. The middle row, figure 8.19b, shows the source estimates after ESD. According to a visual inspection of the source images, the strong global signal is concentrated in the leftmost source and the stripe-like mapping signal in the second source pattern from right. Sources no. 4, 5, and 7 from left contain different blood-vessel patterns and blood-vessel movement artifacts and the remaining sources contain noise. Compared to the input pattern of the top row, the mapping signal could be successfully separated from the other components, and could be concentrated in a single source pattern. The plots in the bottom row result from the back-projection of the respective source pattern (middle) onto the input data set.

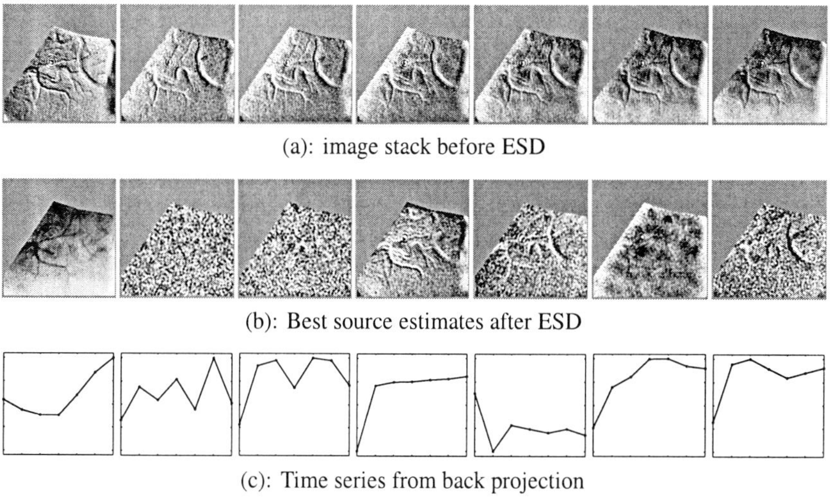


*Figure 8.18.* **(a)** Single-condition stack for left-eye stimulation after first-frame analysis, masking of a region of interest and lowpass-filtering (cutoff: 10.6 $mm^{-1}$, $\lambda = 0.094$ mm). Time from left to right: -1,0,1,2,3,4,5 sec. relative to stimulus onset. **(b)** Spatial patterns of source estimates after ESD analysis. **(c)** Normalized Back-projections of source patterns on the data set in (a). Time axis: $t \in [-1, 5]$.

The mapping signal and the global signal smoothly increase over time, but the time delay of the global signal is bigger than that of the mapping signal. This smooth and stimulus-locked temporal change can be used as an additional hint of the origin of a source pattern. However, global and mapping components must be distinguished by some additional knowledge about their behavior in space (cf. discussion below).

Figure 8.19 summarizes ESD results for all image stacks of an ocular dominance experiment. A standard procedure for the generation of single-condition images involves first frame analysis, temporal summation of the images and lowpass-filtering. The two single-condition images for left- and right-eye stimulation from this procedure are shown in figure 8.19a and 8.19b. The mapping signal is strongly mixed with blood-vessel patterns and spatially global reflectance changes. Note that this corruption cannot be reversed by bandpass filtering, which generated the image of figure 5.9. Figure 8.19c and 8.19d show the resulting mapping source pattern from an individual ESD analysis of the left-eye and right-eye single-condition stack. In both maps, the mapping signal is weak and strongly noisy, but has been reasonably well separated from the other patterns. ESD applied on the OD difference image stack leads to the differential map shown in 8.19e, in which the signal-to noise ratio has been improved compared to pure temporal summation. Figure 8.19f results from

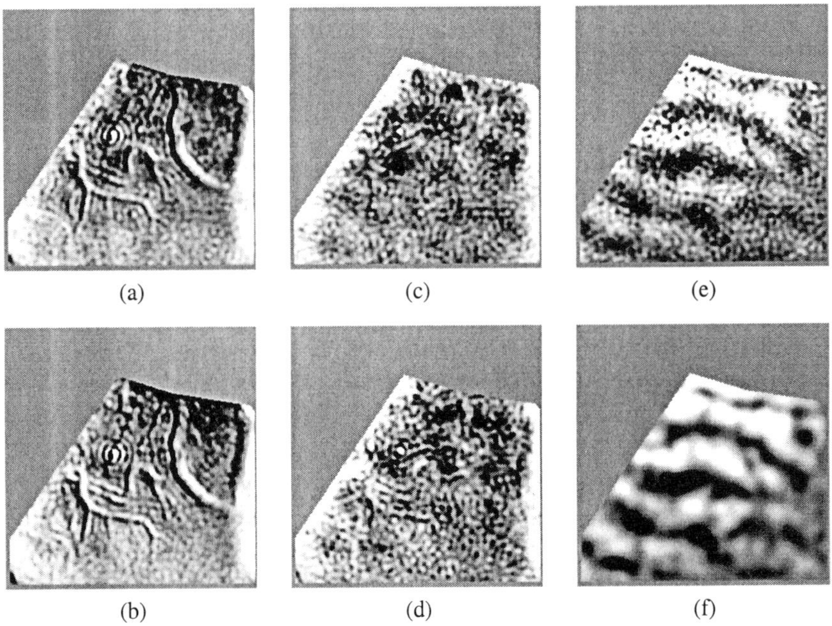

*Figure 8.19.* Left column: Single-condition images obtained by first-frame analysis, lowpass-filtering and temporal summation. **(a)** Left-eye stimulation (cf. figure 8.18a) and **(b)** right-eye stimulation. **(c)** and**(d)** Corresponding single-condition maps for left and right-eye stimulation, respectively, after ESD analysis. **(e)** Ocular dominance map from ESD-analysis of the differential image stack. **(f)** Lowpass-filtered version of (e) (cutoff 4 mm$^{-1}$, $\lambda = 0.25$ mm).

8.19e after lowpass-filtering with a lower cutoff-frequency. Note that filtering should be done after source separation rather than before: Had we filtered the data before with the strong lowpass-filter, the other source patterns such as the blood-vessel patterns would have been strongly blurred and would have more strongly corrupted the mapping signal. After ESD we can be more confident to look at the mapping signal, for which high frequencies components are damped out by light-scattering in tissue (cf chapter 4). This justifies the application of a lowpass-filter to the resulting map.

## 4. Concluding Remarks

We used projection methods for the extraction of the mapping signal of optical recordings from other signal components, which are less closely related to stimulated neuronal activity.

If data are analyzed using bandpass filtering with a given set of cutoff frequencies $f_1$ and $f_2$, this corresponds to the statistical model "signals are concentrated within the frequency band $[f_1, f_2]$, the background is concentrated outside". Thus, every pair of cutoff frequencies serves as parameters for the statistical

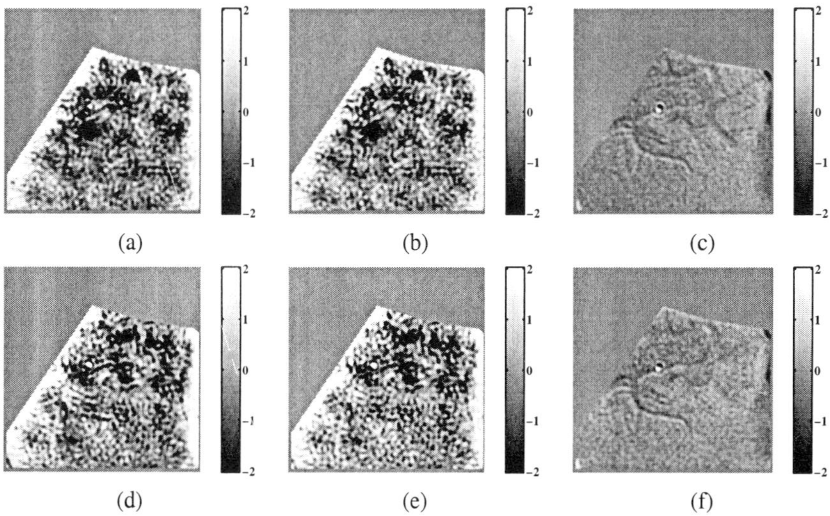

*Figure 8.20.* Top row: **(a)** Left-eye single-condition map after single-shift ESD (vector (5,0)), **(b)** averaged result over ten shift vectors ($\Delta r_x = 1+2i, i = 0, ..., 9$) and **(c)** Difference between both maps. **(d)** - **(e)**: Same for right-eye single condition map. The separation results is fairly insensitive against the particular choice of the shift vector.

model, which can be arbitrarily chosen. In contrast, the methods applied in this chapter are based on parameter-free statistical assumptions, namely "the signals are orthogonal and can be distinguished by variance" (PCA), or "different signal sources are smooth and their cross-correlation functions vanish" (ESD). The shift vector of ESD does not touch the statistical model, i.e. the independence assumption. Rather, the the same kind of independence is assumed for all shift vectors. In particular, this implies that the results obtained for several shifts can be averaged. Figure 8.20 compares the two single-condition maps obtained for a single-shift ESD (8.20a, 8.20d) with the average obtained from ten ESD applications with different shift vector (8.20b, 8.20e). The right column (8.20c and 8.20f) displays the difference images between the single-shift and the averaged map: Both results agree well with each other in that the procedure is relatively insensitive against the detailed choice of the shift-vector. In addition, the difference contains some noise. Hence, averaging can help reducing the noise in the resulting map. However because the sensor noise transformed by different similar unmixing matrices is strongly correlated, noise reduction is not very prominent.

Let us briefly discuss the assumptions on which ESD for optical imaging is based. Though not proven to be true, smoothness over space is very likely to be given for all signal components. For the mapping and global signals, smoothness is generated by the strong scattering of light within the tissue, which blurs all components over space by at least 100-200 $\mu$m (cf chapter 4 and (Stetter and Obermayer, 1999)). For blood-vessels on the cortical surface, which are continuous structures in one dimension, the autocorrelation functions can extend to more than one millimeter. The assumption of vanishing cross-correlations between the sources is less intuitive, yet in the case of orientation maps, where the mapping signal is oscillatory as opposed to the global signal, the resulting cross-correlation functions between the two patterns are small. Also, we observed that cross-correlations between blood vessel patterns, that were extracted from images of the cortex under illumination with green light, and ocular dominance stripes from differential images are less than 20 % of the autocorrelations (Schießl et al., 2000a). Generally the shift vector or vectors for ESD have to be chosen in a range, where both the autocorrelations of the sources and the cross-correlations between the mixtures are large. According to our experience, these conditions are usually fulfilled for arbitrary small shift vectors. ESD can also be applied in a multishift mode (Schöner et al., 2000). If all sets of reasonably small shift vectors are used, multishift ESD does not contain any free parameters anymore. On most of the real data sets, the performance of multishift ESD (data not shown) has turned out to be very similar to that of single-shift ESD, mostly due to the reasons discussed in relation to figure 8.20.

While making use of knowledge about spatial coherence, ESD as applied in this work is confined to second order statistics, because it is based on second-order correlation functions. Though this might appear as a limitation, a recent comparison (Schießl et al., 1999) of ESD with Independent Component Analysis (ICA) based on Infomax (Bell and Sejnowski, 1995) (moments of any order) and on kurtosis optimization (Hyvärinen and Oja, 1997) (fourth order moments, cf. also chapter 7) showed that ESD provides the best separation results for spatially smooth data. This is probably due to the fact that, in the presence of noise, higher order moments are more difficult to estimate using a limited set of data points, than moments of second order. Also, the ICA algorithms usually do not exploit the spatial coherence of patterns as done by ESD.

PCA and ESD are based on fairly general assumptions, but both methods only work if those assumptions are true, and some care has to be taken when interpreting the results. The only direct test for the biological relevance of both differential maps and PCA/ESD maps is a comparison between the optical maps and the corresponding electrophysiologically measured neuronal response. However, an indirect method for scoring the reconstructed maps pops out of the method: If the spatial neuronal activation patterns have already been characterized, the spatial distributions help identifying the mapping signal, when accompanied by a reasonable conjugate time series. If no information about spatial neuronal activation patterns is available beforehand, the shape of the time-series, which must change coherently with the stimulus onset, can serve as a criterion for the identification of the mapping component. In any case, however, the only direct test for the biological relevance of both differential maps and PCA/ESD maps is a comparison between the optical maps and the corresponding electrophysiologically measured neuronal response.

In summary, there is a variety of recently developed statistical procedures, which allow to reliably extract indicator signals for cortical neuronal activity patterns from optical measurements. Once we have access to the spatial response characteristics of cortical tissue, we possess a promising basis for revealing some of the secrets of cortical information representation. Experimental activation data serve as important input to functional models for cortical representation of information, which will be introduced in the following chapters.

aspects of their operation prove similar to experimentally observed cortical neuronal response properties. This demonstrates, that cortical function is compatible with the design principles imposed: it works optimal in the sense formulated in the design principle. This by no means implies that we do now fully understand cortical function. For example, many functional aspects are compatible with several (virtually with every reasonable) design principles and consequently one particular optimality criterion is not sufficient to understand how the brain works. Also, many functional aspects (e.g. temporal coding) are not even addressed so far, but may be at least as important as the spatial code. Finally, computational models show why it might be reasonable for the brain to operate in the particular observed way but give no answer to the question how a particular algorithm is implemented in the cortical anatomical substrate. This latter question is addressed by connectionist models of cortical function, some of which we summarize in the next chapter.

Chapter 10

# MEAN-FIELD MODELING OF CORTICAL FUNCTION

In chapter 3 we have provided a brief overview over the neuroanatomy of the macaque cortex. It has become clear that cortical tissue has a highly complex structure with many neuron types, each of which again shows a considerable diversity across the population of individual nerve cells. By use of mathematical models for cortical function, we hope to obtain a deeper understanding of what is going on in our brain. Computational models (cf. previous chapter) help to understand whether or not a general design principle of signal processing is compatible with observed aspects of brain function. However, computational models rarely deal with the question of how an algorithm is implemented in the brain and therefore might be subject to oversimplification. They leave us with the question how cortical information representation is generated by the network of neurons in cortical tissue: How does cortical function result from cortical structure?

One possibility to address this question is to generate a detailed model of cortex which mirrors the anatomy and physiology as closely as possible. However, modeling every detail of cortical anatomy represents an extremely difficult and tedious if not impossible task. In addition, even if we were able to build a one-to-one model of a piece of cortex, our model would behave exactly as the cortex itself and thereby we would not be able to understand it any better as its biological counterpart. Finally, we might add many aspects to the model which are unique to one particular piece of cortex (say, of one particular animal) and are not general cortical features. In this sense, forming a very detailed model of cortical function would "overfit" the system by explaining the details rather than the general principles.

Connectionist models of cortical function want to extract the general underlying principles for the observed structure-function relationships: Which of the anatomical features observed in biological cortices are important and necessary ingredients for the generation of the observed cortical response properties? Consequently, a connectionist model should act at an intermediate complexity level, and it is one major task for the modeler to find the correct level of complexity. Once good models are found and basic principles of cortical information processing are extracted, our reward might be the key to a "general problem solver" module: Although cortical function is very diverse (from local scene analysis in the primary visual cortex up to reasoning and decision making in the prefrontal cortex (Damasio, 1996)), the anatomy of the cortex is astonishingly uniform over all areas. This observation has led to the hypothesis that the cortex carries out the same (or similar) operations to different types of input. By understanding how the primary visual cortex works, we can therefore hope to derive general construction principles of the whole cortex.

In the following sections, we will address the question how cortical responses (summarized in chapter 3, section 2) are generated by the neuronal circuitry of the early visual system. We will start from simple models and will identify and add more and more necessary ingredients for the observed response behavior.

## 1. Iceberg-Model of Orientation Selectivity

One important early model for orientation tuning first formulated by Hubel and Wiesel, e.g. (Hubel and Wiesel, 1977), is based on the structure of simple receptive field profiles (figures 3.12 and 9.3). They consist of elongated subfields with alternating ON- and OFF-response. The Iceberg-model proposes that orientation selectivity is generated by filtering oriented input with the receptive field profile (calculating its overlap with the profile) and feeding the result through a rectifying nonlinearity. This means that the model assumes that orientation selectivity is purely generated by feed-forward processing of input (mediated by the feed-forward fibers from the retina over the LGN to a cortical neuron) and local processing within the neuron. In particular, no function is assigned to the intracortical circuitry as well as feedback-circuitry to the LGN. These connections are neglected in the iceberg model and it is sufficient to consider the response of an individual cortical neuron.

The iceberg-model for oriented cortical response can be mathematically formulated by a simple neural network, the architecture of which is illustrated in figure 10.1. Two layers of ON- and OFF-center LGN cells process the visual input and project to a single considered cortical neuron. The projections are structured in a way that only ON- and OFF-LGN cells with spatially aligned concentric receptive fields (figure 10.1, right part) drive the cortical simple cell by non-negative weights $w^{\mathrm{ON}}$ and $w^{\mathrm{OFF}}$, respectively. This particular arrangement of afferent LGN cells is viewed as the origin of elongated ON and OFF subfields of simple receptive fields. Because we do not address aspects of temporal coding for the moment, it is sufficient to assume a rate code for the neurons. ON-LGN-cells increase (decrease) their spike rate relative to the

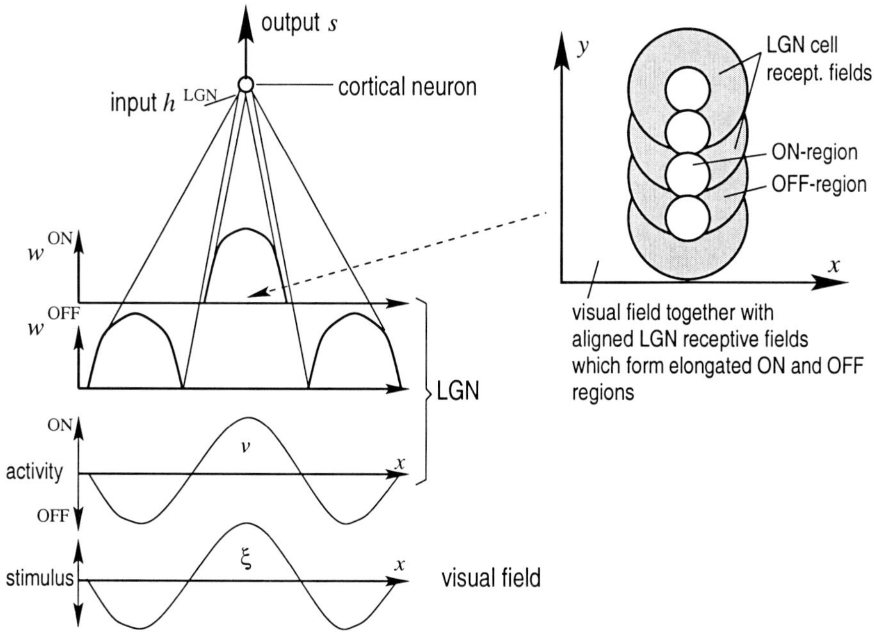


*Figure 10.1.* One-dimensional cross-section through a simple neural network for the iceberg model. A cortical neuron is selectively fed by ON- and OFF-type LGN neurons with aligned receptive field centers. The drawing on the right part illustrates the alignment of ON-LGN receptive fields with nonzero weights $w^{\mathrm{ON}}$ to the cortical neuron. Bottom plots sketch a zero-phase grating stimulus $\xi$ (bottom) and the corresponding effective LGN activity $v$ (top), which are very similar to each other.

spontaneous activity level (which is relatively high for LGN cells), if a bright (dark) light bar covers their center. This deviation is denoted by $v^{\mathrm{ON}}$. The relative change in spike rate for OFF-center LGN cells is denoted by $v^{\mathrm{OFF}}$. Because OFF-center LGN cells show roughly the opposite response behavior to ON-center cells, we can approximately set $v^{\mathrm{ON}} \approx -v^{\mathrm{OFF}} \equiv v$, where $v$ is referred to as effective LGN response. The second plot from bottom in figure 10.1 schematically illustrates the effective response of LGN neurons over space, if a zero-phase grating is presented to the network (bottom plot). Because the activity $v$ results as a linear filtering of the sine-wave input $\xi$ by the Mexican hat like LGN receptive field profile, $v$ and $\xi$ have the same phase and frequency and differ only in amplitude: $v = v_0\xi$. Based on these observations, the total synaptic input of the cortical neuron becomes

$$h^{\mathrm{LGN}} = \sum_a w_a^{\mathrm{ON}} v_a^{\mathrm{ON}} + w_a^{\mathrm{OFF}} v_a^{\mathrm{OFF}} \qquad (10.1)$$

$$\approx \sum_a w_a^{\mathrm{ON}} v_a - w_a^{\mathrm{OFF}} v_a \tag{10.2}$$

$$= v_0 \sum_a (w_a^{\mathrm{ON}} - w_a^{\mathrm{OFF}}) \xi, \tag{10.3}$$

where $a$ runs over all LGN cells which drive the cortical neuron. In the present linear model framework, the total synaptic input of the cortical neuron (up to a constant factor) can be calculated as the overlap between the input pattern and the filter $G(\mathbf{r})$ with $G(\mathbf{r}_a) = w_a^{\mathrm{ON}} - w_a^{\mathrm{OFF}}$. $\mathbf{r}_a$ denotes the receptive field center of the $a$-th LGN cell in the visual field. This filter represents the model of a simple receptive field profiles, which have been found to resemble Gabor-filters (Daugman, 1980).

In the rate model, the output of the cortical neuron can be calculated as a nonlinear transfer function of the input. Motivated by thresholded semilinear current-frequency relationships shown in figure 2.9a, we can identify the input current with the total synaptic input $h^{\mathrm{LGN}}$ and can calculate the output as a thresholded and rectified version of the input:

$$s = f(h^{\mathrm{LGN}}) = \max(h^{\mathrm{LGN}} - T, 0) \tag{10.4}$$

Note that by adopting the behavior of the frequency-current relationship, we implicitly assume stationarity of the system.

Orientation selectivity is generated in the model framework as follows: If the stimulus is an oriented structure such as a sine wave grating, the total synaptic input will be maximal, if the orientation and phase of the sine wave is aligned with orientation and phase of the receptive field profile, and will be zero, if the structures are orthogonal. The two rows at the bottom of figure 10.2 show different oriented sine wave gratings and the Gabor-filter $G(\mathbf{r})$. In the example shown, the overlap is maximal for the middle stimulus and decreases with the difference between orientations. The resulting synaptic input for two different contrast levels of the stimulus grating is plotted in the bottom diagram against the stimulus orientation. The simulation was performed using differently oriented sine-waves as stimuli and a Gabor-filter with $\sigma k = 0.2$ (roughly one oscillation within the non-zero region) as receptive field profile. The neuronal output is shown in the top diagram of figure 10.2, and was obtained using eq. (10.4) and a threshold of $T = 0.4$ (dashed line in plot below). The response is orientation selective, but nevertheless the iceberg model has some serious drawbacks, which motivate the consideration of more sophisticated models for cortical function:

- In biology, the sharpness of orientation tuning is independent of stimulus contrast (cf. figure 3.13a and (Sclar and Freeman, 1982)). This independence might be important, as it ensures that stimulus orientation and stimulus contrast are coded independently of each other in the cortex, which facilitates the independent readout of both quantities. The iceberg-model predicts a strongly contrast-dependent orientation tuning width.

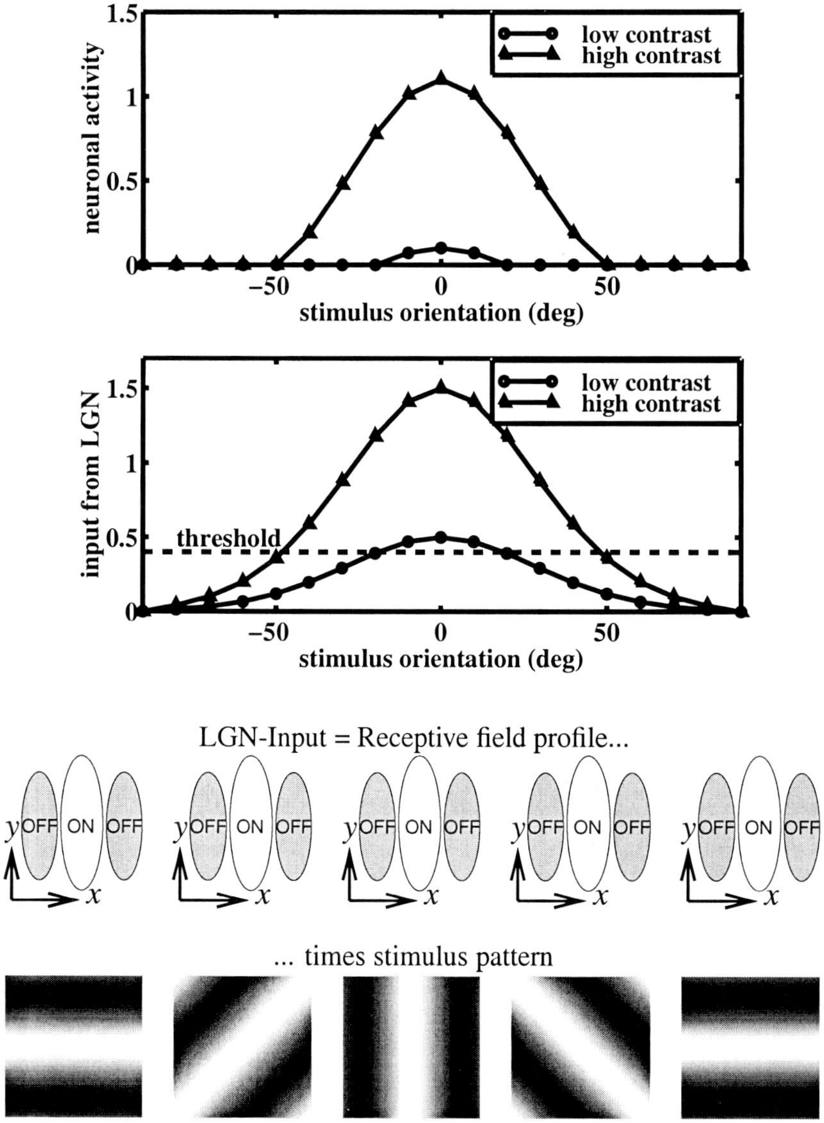


*Figure 10.2.* The iceberg-model of orientation selectivity. Bottom rows: Filtering oriented input by a simple receptive field profile yields an orientation tuned input from the LGN to a cortical neuron (bottom diagram). Top diagram: The output of the cortical neuron is a thresholded version of its input. The output is orientation selective, but its tuning width increases with stimulus contrast.

- In biology the orientation tuning curves are sharper than predicted by the iceberg model, and the output of neurons is often more sharply tuned than their input (Volgushev et al., 2000). If rectifying nonlinearities of LGN cells are taken into account in the model instead of the linear filter model assumed here, its simulated tuning curves become even broader.

- The iceberg model does not assign any function to intracortical circuitry, which in fact is even much stronger than the feed-forward circuitry. This leaves the question of which operations are performed by lateral intracortical connections.

- The iceberg model is not suitable for modeling more complex features of cortical processing. For example it cannot explain, why the neuronal response curves obtained from neuronal stimulation (cf. 2.9b) are completely different from those obtained by electrical stimulation (cf. 2.9a).

In light of these observation, we face the necessity to build a more sophisticated model of cortical function, which involves dense recurrent intracortical circuitry (Somers et al., 1995; Ben-Yishai et al., 1995; Ben-Yishai et al., 1997; Bartsch et al., 1997; Hansel and Sompolinsky, 1998). A sufficiently large cortical circuit may contain millions of densely interconnected neurons, which cannot be modeled at a single neuron level because a prohibitive computational expense. We face the task to formulate a model framework which allows a simplified yet still valuable theoretical description of a large neuron population at a mesoscopic level. One important type of mesoscopic cortical description has been provided by mean-field models of cortical function, which will be introduced in the next section.

## 2. Mean-Field Model of Neuronal Population Activity

Modeling of cortical signal processing can be considerably simplified by taking into account the columnar structure of the cortex. Each cortical column contains many thousands of neurons, which receive approximately the same afferent and intracortical input and show similar response properties. But even if some response properties (such as direction selectivity in macaque V1) change over depth, there is still a topographic mapping of functional aspects: In most of the cases many nearby neurons show similar selectivities.

### 2.1 Principle and Basic Assumptions

Based on these observations, (Ben-Yishai et al., 1995) and (Bartsch et al., 1997) have formulated a mean-field description of cortical processing. First, if two neurons $k$ and $l$ within a cortical column receive similar total synaptic inputs, $h_k \approx h_l$, it seems a good approximation to assume that each neuron within a given population $\alpha$, for example all excitatory ($\alpha$ = "$e$") or all inhibitory

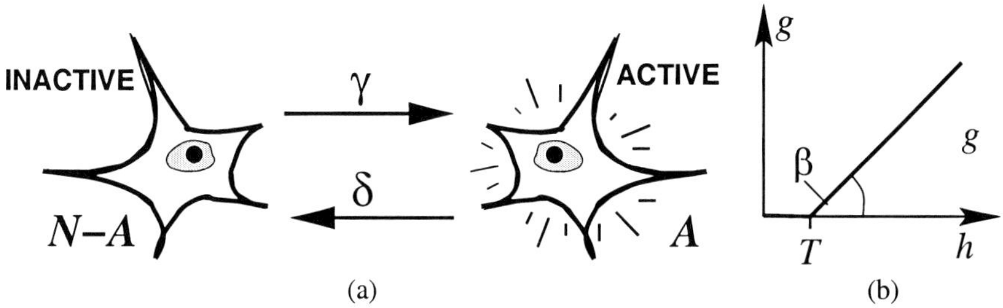


*Figure 10.3.* **(a)** Principle of binary stochastic neurons: The neuron randomly toggles forth and back between an active (right) and inactive (left) state. If the neuron is driven by excitatory input, the probability for activation increases and inactivation probability decreases resulting in a higher mean activity. **(b)** Semilinear activation function $g(h)$.

neurons ($\alpha =$ "$i''$") within a column, receive the same input, $h_{k,\alpha} \equiv h_\alpha$ for all neurons $k$ of the population. In other words, instead of feeding to every neuron its actual input, all neurons within one population are assumed to receive the same mean input. Due to that reason, this kind of models is referred to as *mean-field models* [1]. Second, within each population many neurons encode similar input properties. Therefore it seems reasonable to hypothesize, that only the overall activity of a whole neuron population is important, whereas the fluctuations within the activities of individual neurons, which can be viewed as nonlinear stochastic units, can be neglected at this level. This corresponds to the assumption that important stimulus properties are encoded in population activities (for aspects of population coding see (Paradiso, 1988; Zohary, 1992)). Note that omitting random fluctuations does not imply that we exclude non-random collective phenomena, i.e. the mean population activity can still be strongly time-dependent. For a model that explicitly includes fluctuations we refer to (Tsodyks and Sejnowski, 1995)). Though neurons of one cell type within a cortical column are natural candidates for a neuron population, the framework can be applied in a more general way: Arbitrary sets of neurons with similar inputs and responses can be combined to a population.

## 2.2 Dynamics of a Neuron Population

We are now ready to write down a simplified description of neuronal population dynamics. One possibility for such a description assumes individual neurons as stochastic neurons binary stochastic units, which can flip forth and back between an active (spiking) and inactive (non-spiking) state. The prob-

[1]This term has been borrowed from solid state physics, where mean-field models describe the behavior of atomic magnetic moments under the mean magnetic fields of their neighbors instead of the actual fluctuating magnetic field.

ability per unit time for an inactive neuron to be activated is denoted by the activation rate $\gamma$ and its inactivation rate by $\delta$ (10.3a). Below, we will relate $\gamma$ and $\delta$ to the mean synaptic input of the population: A high excitatory input will lead to a high activation rate and a low inactivation rate, which causes a higher fraction of neurons to be active. Low input or inhibitory input, in contrast will reduce the activation rate and the neurons settle down to the inactive state. If we consider a population of $N$ binary stochastic neurons with identical activation and inactivation rates, the number of active neurons $A$ changes within the small time interval $\Delta t$ according to

$$\Delta A = \gamma \Delta t(N - A) - \delta \Delta t A \; = \; (\gamma N - (\gamma + \delta)A) \; \Delta t. \tag{10.5}$$

In the limit $\Delta t \to 0$, this relation transforms to a rate equation for the fraction of neurons $m = A/N$ that are active at time $t$:

$$\frac{d}{dt} m = \gamma - (\gamma + \delta) m. \tag{10.6}$$

Now we assume for simplicity, that the inactivation rate behaves inversely to the activation rate. This is reasonable because neurons which are strongly driven by input are less likely to stop firing spontaneously. If $\gamma_{\text{max}}$ denotes the maximum possible activation rate, we arrive at $\delta = \gamma_{\text{max}} - \gamma$ and

$$\frac{d}{dt} m = \gamma - \gamma_{\text{max}} m. \tag{10.7}$$

One possible interpretation of the maximum activation is related to the refractory period of biological neurons. For a neuron to undergo two subsequent activations, it must at least fire a spike and wait for the refractory period $\tau$ until it can be activated to fire the next spike. This means that we can identify the maximum activation rate with $\gamma_{\text{max}} = 1/\tau$. The rate equation for the population activity becomes

$$\tau \frac{d}{dt} m = -m + \tau\gamma =: -m + g, \tag{10.8}$$

where $0 \le g = \tau\gamma \le 1$ denotes the activation probability (relative to $\tau$) for the neuron population.

For realistic regimes of operation, eq (10.8) can be interpreted as the dynamics of a pool of spiking neurons. This view can be motivated as follows: The (absolute) refractory period $\tau$ is in the range of 1-2 ms, which means that electrically driven neurons can reach spike rates of approximately $500 - 1000$ Hz. In contrast, the spike rates observed for visually stimulated cortical neurons range around 50 Hz: realistic activation rates are much smaller than the maximal rate, $\gamma \ll \gamma_{\text{max}}$, ($g \ll 1$), and consequently the inactivation rate is very fast: $\delta \approx \gamma_{\text{max}} = 1/\tau$. In this regime, each time a neuron becomes activated and fires a spike, it immediately becomes inactivated again with a rate close to its refractory period. In other words, a binary stochastic neuron which is activated fires an individual spike and automatically inactivates again: In the limit of low mean activation, our mean-field model describes a population of spiking (instead of binary) neurons.

Now we have to specify, how the activation probability changes with the synaptic input. If the input of a neuron population falls below a threshold $T$, all neurons of the population are inactive, otherwise, neurons should be activated the more frequently, the more excitatory input they receive. The simplest function, which preserves this important rectifying nonlinearity present in biological neuronal systems is a semilinear function, and we can formulate the dependence of the activation probability $g$ on the mean synaptic input $h$ of the population as

$$g(h) = \max(\beta(h - T), 0). \tag{10.9}$$

$g(h)$ is referred to as the activation function of the population (figure 10.3b). Note that at this point we have made use of the mean-field assumption. In eq. (10.9), $h$ has to be the mean synaptic input of the population. If we had used the actual synaptic input, the activation function would have different values for each neuron in the population and the ensemble average could not be as easily written down as in eq. (10.8).

## 3. Modeling Orientation Selectivity with two Cell Types

In chapter 3, section 2.4, we have seen that the primary visual cortex processes visual input locally: Each local visual feature is processed and represented by a patch of cortex about $1 - 2$ mm in diameter, which is called a hypercolumn. The neurons within a hypercolumn are densely connected by lateral intracortical fibers and form a strongly coupled recurrent network. In contrast to the considerations of the iceberg model, we wish to explore, how the afferent input (transformed visual signals) is processed by this recurrent cortical network. We ask which signal processing operations are implemented in the recurrent network. By formulating a mean-field model of a cortical hypercolumn (Ben-Yishai et al., 1995; Ben-Yishai et al., 1997; Bartsch et al., 1997; Bartsch et al., 1999b; Stetter et al., 2000a), we can efficiently describe the dynamics of hundreds of thousands of neurons within a hypercolumn at a population level.

### 3.1 Model Setup for a Hypercolumn

A model hypercolumn for the processing of oriented stimuli as illustrated in figure 10.4 can be constructed by considering the following observations: $(i)$ The primary visual cortex has a columnar structure, $(ii)$ within each column many neurons respond approximately to the same stimulus orientation, and $(iii)$

$$+\underbrace{c\ \{1-\varepsilon+\varepsilon\cos\left[2(\theta-\theta_0)\right]\}}_{h^{\mathrm{LGN}}}\quad \alpha,\beta=e,i \quad (10.17)$$

Figure 10.4c shows a typical choice for the profiles of the lateral coupling strengths. Excitatory connections are mediated by local collaterals of pyramidal neurons, which are strongest between neurons with similar orientation preferences and almost zero for pairs with orthogonal orientation preferences (top curve in figure 10.4c). Note that the fibers are localized in orientation space but need not necessarily be localized in spatial cortical coordinates. For example, within the $2 \times 2$ mm area of the hypercolumn, the same orientation preference appears several times (cf. figures 3.16 and 8.15) periodically with the typical wavelength of the orientation map (Müller et al., 2000). Iso-orientation specific lateral connectivity would then form a set of patches of connectivity over the iso-orientation regions (which has been actually observed by (Malach et al., 1993), see also chapter 3, section 2.4). Inhibitory connections are often assumed to spread further laterally in orientation space than excitatory fibers, i.e. there is non-zero cross-orientation inhibition (bottom curve in figure 10.4c). These longer range lateral inhibition can be assigned to basket neurons (Lund, 1987), which form relatively large axonal arbors. For the parameter choice shown, the fibers do not distinguish between the types of target neurons, i.e. $S_{e,e} = S_{i,e}$ and $S_{e,i} = S_{i,i}$.

In the following sections, we will adopt this anatomically motivated connection scheme. In addition, the activation function of the inhibitory neuron population is assumed to have a higher threshold ($T_i > T_e$) and a higher gain ($\beta_i > \beta_e$) than that of the excitatory population, as reported by (McCormick et al., 1985).

## 3.2 Simulating Contrast-Invariant Orientation Tuning

Figure 10.5 shows, how a hypercolumn of recurrently connected cortical orientation columns represents an oriented stimulus. The presence of an oriented contour or grating within the aggregate field of the hypercolumn evokes a broadly tuned input, which is shown in figure 10.5a for three log contrast levels. The cortical activities are iterated through the recurrent circuitry, until a stationary state is reached. The resulting activity pattern over the 42 orientation columns and for the three contrast levels are shown in figure 10.5b. The activity pattern is more sharply tuned in orientation space than the input, and its tuning width is independent of contrast as observed in biological systems. Note the symmetry between the index $\theta$ of the orientation column and the stimulus orientation $\theta_0$, $h^{\mathrm{LGN}}(\theta - \theta_0) = h^{\mathrm{LGN}}(\theta_0 - \theta)$. As a consequence, the activity pattern over the whole hypercolumn for a single stimulus orientation is the same as the response of one orientation column measured over all stimulus orientations. In other words, the functional form of the activity pattern in the

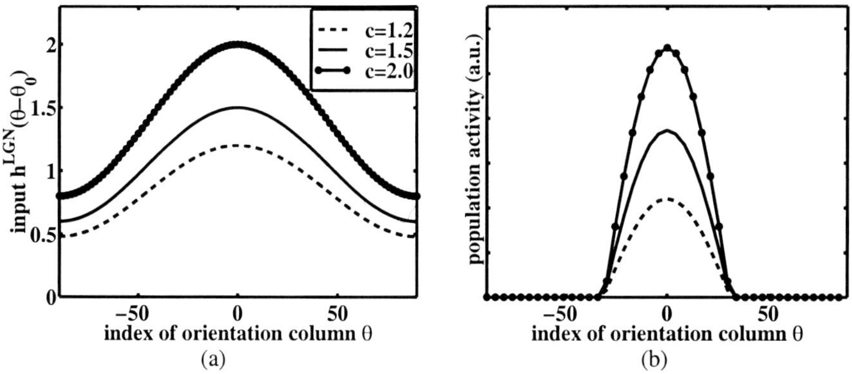


*Figure 10.5.* **(a)** Simulation of orientation representation in a hypercolumn with two types of neurons. **(a)** Weakly tuned input $h^{\mathrm{LGN}}(\theta)$ to the hypercolumn for $\theta_0 = 0$ deg stimulation with three log contrast levels. **(b)** Mean activities of the excitatory neurons of the orientation columns in the stationary state for stimulation as shown in (a). The activity pattern is sharply tuned and its width is independent of contrast. Parameters: 42 columns, $S_{ee} = S_{ie} = S_{ei} = S_{ii} = 30$; $T_e = 1, T_i = 2, \beta_e = 0.5, \beta_i = 1, \sigma_{ee} = \sigma_{ie} = 34$ deg, $\sigma_{ei} = \sigma_{ii} = \infty$.

model is identical to the orientation tuning curve of one column. Therefore, we will henceforth identify both quantities with each other and refer to plots as in figure 10.5b as orientation tuning curves. In summary, a model hypercolumn with strong recurrent and local excitation and a distributed recurrent inhibition represents the stimulus orientation independently of contrast. This regime of operation of a hypercolumn has been referred to as "marginal phase" (Ben-Yishai et al., 1995), because the cortical representation is preferentially determined by intracortical dynamics instead of the detailed characteristics of afferent input.

## 3.3 Analytical Treatment of Orientation Tuning

In order to gather an analytical understanding of the representation of oriented stimuli, we consider the cosine-shaped connectivity patterns of eqs. (10.14) and (10.15) with $E_0 \geq E_2 \geq 0$ and $I_0 \geq I_2 \geq 0$. Further, we denote by $M_\alpha(\theta)$ the steady state activity of the population $\alpha$, which arises as the stationary solution of eq (10.16) and define its zeroth order and the second order Fourier coefficients

$$m_{\alpha,0} := \int \mathrm{d}\theta'/\pi \; M_\alpha(\theta'), \tag{10.18}$$

$$m_{\alpha,2} := \int \mathrm{d}\theta'/\pi \; M_\alpha(\theta') \cos(2\theta'). \tag{10.19}$$

It can be shown (Bartsch et al., 1997; Stetter et al., 2000a), that in the stationary state $m_{\alpha,0}$ and $m_{\alpha,2}$ follow the conditions

$$m_{\alpha,0} = \frac{\beta_\alpha}{\pi}(\varepsilon c + E_2 m_{e,2} - I_2 m_{i,2})(\sin(2\theta_{c,\alpha}) - 2\theta_{c,\alpha}\cos(2\theta_{c,\alpha})) \quad (10.20)$$

$$m_{\alpha,2} = \frac{\beta_\alpha}{\pi}(\varepsilon c + E_2 m_{e,2} - I_2 m_{i,2})\,(\theta_{c,\alpha} - \sin(4\theta_{c,\alpha})/4)\,, \quad (10.21)$$

where $\theta_{c,\alpha}$ is the tuning width (measured at the basis of the tuning curve) of population $\alpha$.

For sufficiently high recurrent excitation, the hypercolumn operates in the marginal phase (Ben-Yishai et al., 1995), in which even a vanishingly small afferent orientation bias is amplified and gives rise to an activity blob whose width is independent of stimulus contrast. For $T_i > T_e$, the width of the blob across inhibitory neurons is slightly smaller than the width of the blob across the excitatory population and both widths depend on stimulus contrast albeit less than 10 percent for $T_i = 2T_e$. These results can be derived as follows. The marginal phase is defined by the stability of an oscillatory solution of the stationary activities $M_\alpha(\theta)$ in the orientation space even for (almost) unbiased input. An oscillatory solution is indicated by the stability of a nonzero value of $m_{\alpha,2}$, which requires the determinant of the right hand side of the linear equation system in $m_{e,2}$ and $m_{i,2}$, eq. (10.21), to vanish. We obtain the condition

$$\beta_e E_2\,(\theta_{c,e} - \sin(4\theta_{c,e})/4) - \beta_i I_2\,(\theta_{c,i} - \sin(4\theta_{c,i})/4) = \pi, \quad (10.22)$$

which for the special case of a flat profile of the lateral inhibition, i.e. $I_2 = 0$, reduces to the expression

$$\beta_e E_2\,(\theta_{c,e} - \sin(4\theta_{c,e})/4) \equiv \beta_e E_2 \pi f(\theta_{c,e}) = \pi, \quad (10.23)$$

or

$$\frac{1}{\beta_e E_2} = f(\theta_{c,e}) \quad (10.24)$$

with $f(x) = (1/\pi)(x - sin(4x)/4)$.

Figure 10.6 illustrates, under which conditions eq. (10.24) can be fulfilled. For $0 \leq \theta \leq \pi/2$, $f$ assumes values between 0 and $1/2$. A solution of equation (10.24) exists only for a sufficiently strong spatial modulation $E_2$ of the lateral excitation profile and thus for a sufficiently strong overall connection strength $E_0 \geq E_2 > 2/\beta_e$. In this case, the orientation tuning width is uniquely determined by the properties of the lateral circuitry and in particular is independent of the stimulus contrast. According to the second term on the left hand side of eq. 10.22, modulation $I_2 > 0$ of lateral inhibition shifts the boundary of the marginal phase to higher values of $E_2$, yet sharpening occurs as long as its modulation over space is weaker than that of the excitatory connections, $I_2 < E_2$.

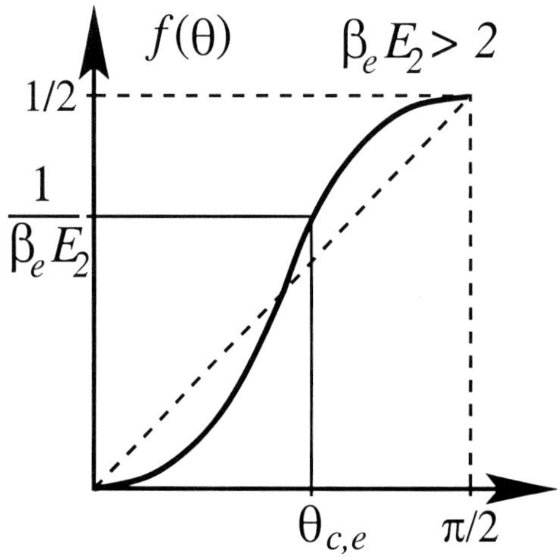


*Figure 10.6* Graphical solution of the condition eq. (10.24). A solution exists only for $\beta_e E_2 > 2$. In this case, the orientation tuning width $\theta_{c,e}$ is uniquely determined and is independent of contrast.

In summary, there is a phase boundary for the strength of the lateral excitation at $\beta_e E_2 = 2$. Above this phase boundary, stimulus orientation is represented by a sharp tuning curve, independently of the strength of afferent orientation bias and independently of stimulus contrast. This holds, as long as the profile of excitatory circuitry is more strongly modulated in orientation space than the inhibitory circuitry profile.

## 4. Modeling Contrast Saturation with two Cell Types

Figure 2.9 has demonstrated that the spike rates of cortical neurons saturate with increasing logarithmic contrast of a visual stimulus (2.9b), but do not saturate, if they are stimulated by a stationary injected current (2.9a). Thus, it seems reasonable to hypothesize, that contrast saturation is an emergent property of the cortical neuronal system instead of an intrinsic property of individual neurons. In this section we address the question, if and under which conditions contrast saturation can be a consequence of the neuronal circuitry (as opposed to synaptic properties (Stetter et al., 1998a)). Again we start with a simple model architecture, in which all excitatory and inhibitory neuron types are lumped together into only two neuron populations 'e' and 'i' per orientation column.

### 4.1 Analytical Treatment of Isolated OR-Columns

First, we wish to analytically describe the basic principles of recurrent signal processing which can serve as the origin of contrast saturation. For this, it has proven useful (Bartsch et al., 1997; Stetter et al., 2000a) to consider the strongly idealized case, in which the different orientation columns are decoupled from each other and only connections within each orientation column are preserved (figure 10.7), i.e., $S_{\alpha\beta}(\theta - \theta') = \text{sign}_\beta S_{\alpha\beta}\delta(\theta - \theta')$. In this case, the average

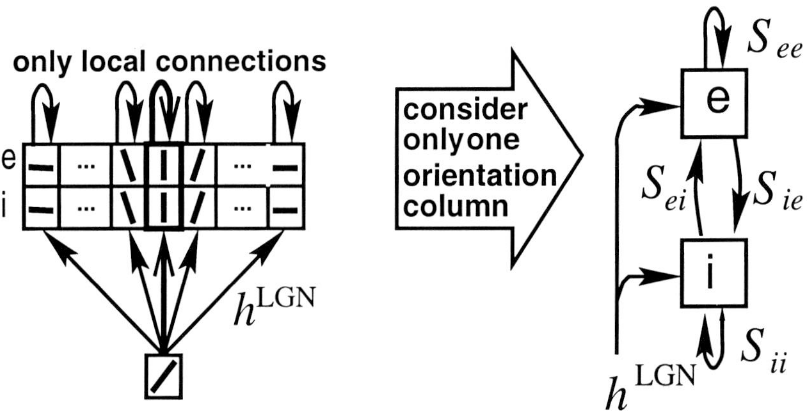


*Figure 10.7.* Strongly idealized version of a hypercolumn, in which only local connections within each orientation column are preserved.

activities of each orientation column evolve independently according to

$$\frac{d}{dt}m_e = -m_e + g_e\left(S_{ee}m_e - S_{ei}m_i + h^{\mathrm{LGN}}\right), \tag{10.25}$$

$$\frac{d}{dt}m_i = -m_i + g_i\left(S_{ie}m_e - S_{ii}m_i + h^{\mathrm{LGN}}\right). \tag{10.26}$$

Also, because the orientation columns are decoupled, only one column needs to be considered, for which we omit the arbitrary orientation index $\theta$.

The stationary solutions $M_e$ and $M_i$ of the coupled equations (10.25) and (10.26) obey

$$M_e = g_e\left(S_{ee}M_e - S_{ei}M_i + h^{\mathrm{LGN}}\right), \tag{10.27}$$

$$M_i = g_i\left(S_{ie}M_e - S_{ii}M_i + h^{\mathrm{LGN}}\right), \tag{10.28}$$

and they can be determined by considering three different intervals of input strengths. For the trivial case of weak sub-threshold input, $h^{\mathrm{LGN}} < T_e \leq T_i$, both activation functions on the right hand sides of eqs (10.27) and (10.28) vanish, which yields zero stationary activations $M_e = M_i = 0$. If the LGN-input is increased slightly above the activation threshold of the excitatory population, this population becomes active, but its activity is not strong enough to excite the inhibitory population also. In this regime, one finds $g_i = 0$ and therefore $M_i = 0$, while the fixed point equation for $M_e$ becomes

$$M_e = \beta_e(S_{ee}M_e + h^{\mathrm{LGN}} - T_e). \tag{10.29}$$

For $\beta_e S_{ee} < 1$ we obtain

$$M_e(h^{\mathrm{LGN}}) = \frac{\beta_e}{1 - \beta_e S_{ee}}(h^{\mathrm{LGN}} - T_e), \quad M_i = 0, \tag{10.30}$$

If the input from the LGN is further increased, the inhibitory population becomes active as well. This happens as soon as the total input of the inhibitory population exceeds its threshold $T_i$. The critical external input $h^{\mathrm{LGN}}_{\mathrm{crit}}$, for which this happens, and which is therefore the contrast threshold for the inhibitory neurons, is then defined by the condition

$$T_i = S_{ie} M_e(h^{\mathrm{LGN}}_{\mathrm{crit}}) + h^{\mathrm{LGN}}_{\mathrm{crit}} = \frac{\beta_e S_{ie}}{1 - \beta_e S_{ee}}(h^{\mathrm{LGN}}_{\mathrm{crit}} - T_e) + h^{\mathrm{LGN}}_{\mathrm{crit}}. \tag{10.31}$$

Solving this equation for the contrast threshold leads to

$$h^{\mathrm{LGN}}_{\mathrm{crit}} = \frac{\beta_e S_{ie} T_e + (1 - \beta_e S_{ee}) T_i}{1 + \beta_e (S_{ie} - S_{ee})}, \tag{10.32}$$

where a nonzero denominator is assured for $\beta_e S_{ee} < 1$.

In the remaining case of $h_{\mathrm{LGN}} > h^{\mathrm{LGN}}_{\mathrm{crit}}$, both activation functions $g_e$ and $g_i$ in eqs. (10.27) and (10.28) can be replaced by their linear parts. The fixed point equation for the system becomes

$$\begin{pmatrix} M_e \\ M_i \end{pmatrix} = \begin{pmatrix} 1 - \beta_e S_{ee} & \beta_e S_{ei} \\ -\beta_i S_{ie} & 1 + \beta_i S_{ii} \end{pmatrix}^{-1} \cdot \begin{pmatrix} \beta_e (h^{\mathrm{LGN}} - T_e) \\ \beta_i (h^{\mathrm{LGN}} - T_i) \end{pmatrix}. \tag{10.33}$$

Inversion of the matrix on the right hand side of eq. (10.33) provides the fixed-point activations

$$\begin{aligned} M_e(h^{\mathrm{LGN}}) &= \frac{(1 + \beta_i (S_{ii} - S_{ei}))}{D} \beta_e (h^{\mathrm{LGN}} - h^{\mathrm{LGN}}_{\mathrm{crit}}) \\ &\quad + M_e(h^{\mathrm{LGN}}_{\mathrm{crit}}) \qquad (10.34) \\ M_i(h^{\mathrm{LGN}}) &= \frac{(1 + \beta_i (S_{ee} - S_{ie}))}{D} \beta_i (h^{\mathrm{LGN}} - h^{\mathrm{LGN}}_{\mathrm{crit}}) \qquad (10.35) \end{aligned}$$

where the denominator is given by $D = 1 - \beta_e S_{ee} + \beta_i S_{ii} + \beta_e \beta_i (S_{ie} S_{ei} - S_{ee} S_{ii})$ and $M_e(h^{\mathrm{LGN}}_{\mathrm{crit}})$ is determined from eqs. (10.32) and (10.30).

These analytical results are summarized once more in figure 10.8. For $T_e < h^{\mathrm{LGN}} \leq h^{\mathrm{LGN}}_{\mathrm{crit}}$, only the excitatory population becomes active and each orientation column behaves like a linear amplifier. Its gain is given by $\beta_e/(1 - \beta_e S_{ee})$ (eq. (10.30)) and diverges as $S_{ee}$ approaches the critical connection strength $S_{ee} = 1/\beta_e$. In the following we will call the parameter regime $0 \leq S_{ee} < 1/\beta_e$ the "linear phase" of the cortical column. If the input exceeds $h^{\mathrm{crit}}_{\mathrm{LGN}}$, the inhibitory neurons become active, suppress the excitatory neurons

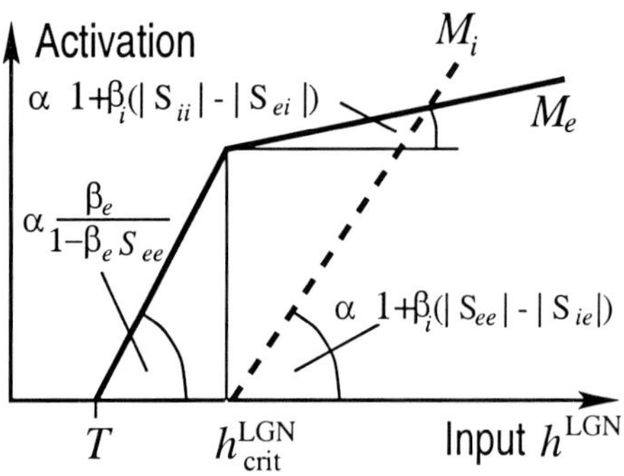


*Figure 10.8* Contrast response of the excitatory ($M_e$, solid) and inhibitory ($M_i$, dashed) population of an isolated orientation column. For weak recurrent excitation, $\beta_e S_{ee} < 1$, the excitatory neurons show contrast saturation as a network effect. The inhibitory neurons do not saturate.

and cause them to saturate. The contrast gain of the excitatory population becomes $(1 + \beta_i(S_{ii} - S_{ei}))/D$ (eq. (10.34)), which may become small, if the excitatory population is inhibited more effectively than the inhibitory population, i.e. $S_{ei} > S_{ii}$. At

$$S_{ei} = S_{ii} + \frac{1}{\beta_i} \tag{10.36}$$

the gain of the excitatory population becomes zero for inputs stronger than $h^{\mathrm{LGN}}_{\mathrm{crit}}$, and for even higher values of $S_{ei}$ supersaturation occurs.

We conclude that in the linear phase, $S_{ee} < 1/\beta_e$, the contrast response curve saturates as the result of the activation of inhibitory neurons with a high activation threshold. The model predicts that their contrast threshold, $h^{\mathrm{LGN}}_{\mathrm{crit}}$, coincides with the stimulus contrast at which excitatory neurons begin to saturate. The gain of the excitatory population at high contrast levels, $h^{\mathrm{LGN}} > h^{\mathrm{LGN}}_{\mathrm{crit}}$, decreases proportional to the difference $S_{ii} - S_{ei}$ between self-inhibition, $S_{ii}$, and inhibition $S_{ei}$ of excitatory neurons. The model also predicts that the inhibitory population does not saturate with stimulus contrast (below the trivial case $M_i = 1$).

## 4.2 Contrast Saturation in the Marginal Phase

Now we consider a hypercolumn with recurrent connections according to eqs. (10.14) and (10.15), which operates in its marginal phase. For this lateral connection scheme it can be shown that the contrast response curve cannot saturate in the marginal phase. For $I_2 = 0$ (flat inhibition) and $E_0 = E_2 = I_0 \equiv S$, we can obtain an analytical expression for $m_{e,2}$:

$$m_{e,2} = \frac{c - T_e}{S^2\beta_i\, d(2\theta_{c,i})/\pi \quad - \quad S^2\beta_e\, d(2\theta_{c,e})/\pi \quad - \quad S\cos(2\theta_{c,e})}, \tag{10.37}$$

with $d(x) = \sin(x) - x\cos(x)$. Because $0 < d(x) \leq \pi$ for $0 < x \leq \pi$, the second Fourier coefficient $m_{e,2}$ is bounded from below by

$$m_{e,2} \geq \frac{c - T_e}{S^2\beta_i + S}. \tag{10.38}$$

As a consequence, $m_{e,2}$ and therefore also the mean activation increases at least proportional with log-contrast $c$, and saturation cannot occur.

## 4.3 Contrast Saturation and Orientation Tuning

The analytical results of limiting cases shown so far hint towards a situation, in which a hypercolumn with two neuron populations per orientation column shows contrast saturation for weak recurrent excitation and contrast-invariant orientation tuning for strong recurrent excitation. In this subsection we study the issue of co-occurrence of contrast saturation and invariant orientation tuning numerically. Three connection patterns of a hypercolumn are considered, namely: $(i)$ Decoupled orientation columns as described in section 4.1 (referred to as "local"), $(ii)$ a cosine-shaped excitation profile balanced by an orientation-independent inhibition as described in section 3.3 (referred to as "cosine/flat"), and $(iii)$ local excitation which is balanced by flat inhibition, i.e. Gaussian profiles eq. (10.13) with $\sigma_{ee} = \sigma_{ie} = 1$ deg, $\sigma_{ii} = \sigma_{ei} = \infty$ (referred to as "local/flat").

Figure 10.9 shows numerical simulations of a hypercolumn which operates in the linear (left column) and in the marginal phase (right column). The top row displays contrast response curves of the hypercolumn, the middle row its orientation tuning curves, and the bottom row the contrast-dependence of the orientation tuning widths. Solid lines, solid lines with circles, and dashed lines correspond to the three connection patterns $(i)$, $(ii)$ and $(iii)$ specified above. The plots demonstrate that in the linear phase contrast response curves saturate (figure 10.9a) but that orientation tuning remains weak and is dependent on contrast (figure 10.9c, 10.9e). In the marginal phase, orientation tuning is sharper and independent of contrast (figure 10.9d, 10.9f), but the contrast response is linear after a sharp onset and does not saturate (figure 10.9b).

Two further observations are worth mentioning: First, the onset of orientation sharpening and the onset of contrast saturation coincide in the linear regime (figure 10.9a, 10.9e). This coincidence is a strong prediction of the model and can be used to test experimentally, whether groups of neurons in the primary visual cortex indeed operate in the linear regime. Second, figure 10.9a shows that saturation of the contrast response curve is weaker in a hypercolumn with flat inhibition compared to a hypercolumn with decoupled orientation columns for the same values of the integrals $S_{\alpha\beta}$ over the connection strengths. For flat inhibition, a large fraction of inhibitory connections received by active neurons originate from cross-oriented neurons, which are silent because of the localized

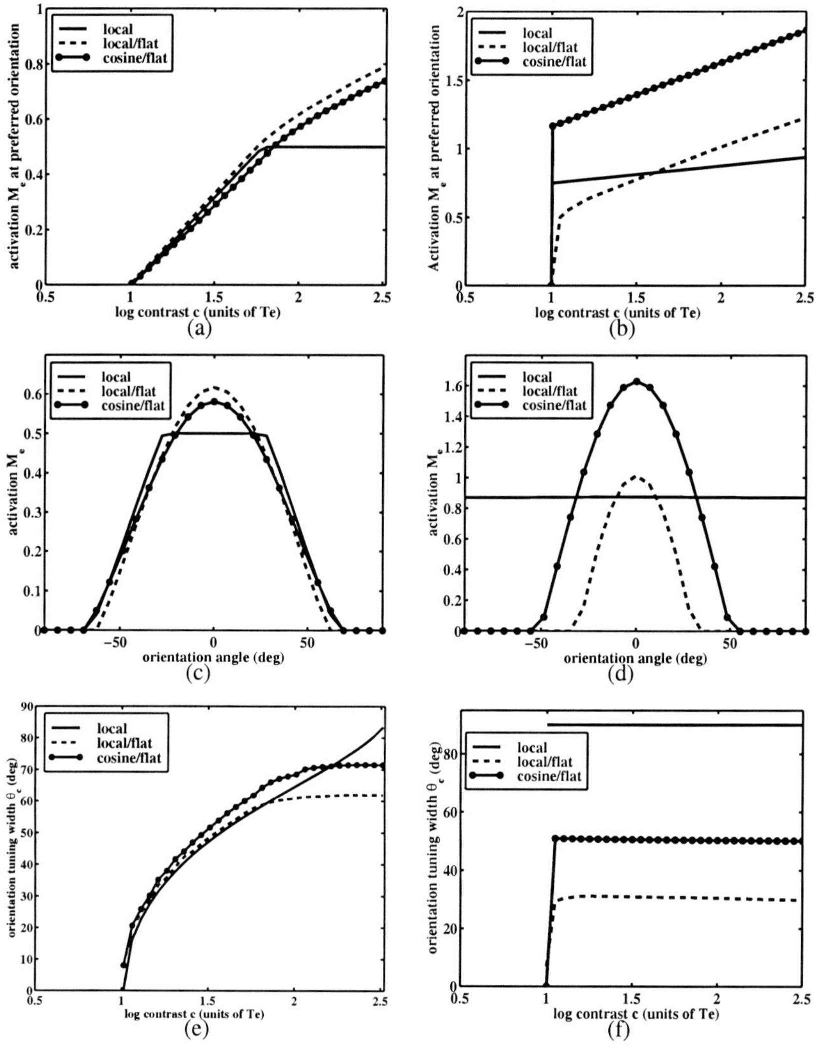


*Figure 10.9.* **(a), (b)** Contrast response curves, **(c), (d)**, orientation tuning curves, and **(e), (f)** the orientation tuning width $\theta_{c,e}$ (half width at the base of the curves in (c), (d)) as a function of log contrast $c$ for the excitatory neuron population. Left column: network operates in the linear phase ($S_{ee} = S_{ie} = S_{ii} = 1,\ S_{ei} = 2$ broadly tuned inputs with $\varepsilon = 0.3$); Right column: network operates in the marginal phase ($S_{ee} = S_{ie} = S_{ei} = S_{ii} = 6$, nearly untuned input $\varepsilon = 0.01$.). Solid lines: "local" couplings. Dashed lines: "local/flat" couplings. Solid circles: "cosine/flat" couplings. Contrast response saturates in the linear phase (a) but not in the marginal phase (b). In the linear phase orientation sharpening is weak (c) and depends on contrast (e), whereas it is sharper and contrast-independent in the marginal phase (d,f). Other parameters were $\beta_e = 0.5, \beta_i = 1, T_e = 1, T_i = 2$, and $c = 2.0$ (in (c), (d)).

activation patterns (figure 10.9c). This renders flat inhibition less effective than local inhibition of the same integral strength. If $S_{ei}$ were increased, however, the hypercolumns with distributed inhibition would also show complete saturation or supersaturation.

Next, the phase boundaries of the linear and marginal regimes were determined numerically for the connection pattern $(iii)$, which may be closest to wiring patterns in area 17. Because the linear phase is characterized by a finite slope of the contrast-response curve close to the activation threshold, $c \simeq T_e$, we calculated the contrast gain (the slope of the contrast response curve) at $c = T_e$. Its divergence marks the boundary of the linear regime. The marginal phase is characterized by the generation of a narrow activation blob from initially untuned input. We stimulated a hypercolumn with weakly tuned input ($\varepsilon = 0.01$) and calculated the resulting orientation tuning width $\theta_{c,e}$. A decrease of the orientation tuning width below 90 deg marks the boundary of the marginal phase.

Figure 10.10 shows the behavior of the contrast gain at the threshold (solid / circles) and the orientation tuning width (crosses) as a function of the connection strength $S \equiv E_0 = E_2 = I_0$. The dotted vertical lines mark the boundary for the linear phase as derived for connection pattern $(i)$ in section 4.1 and for the marginal phase as derived for connection pattern $(iii)$ in section 3.3. Analytical and numerical values for the phase boundaries agree well with each other and the simulations demonstrate that the linear and marginal phases do not overlap. For the boundary of the linear phase, the close correspondence of analytical and numerical values is due to the fact, that in the simulation the input was weakly tuned and evoked a flat activation pattern. All orientation columns have similar average activities and can therefore be approximated by a single orientation column. For the boundary of the marginal phase, the small deviation between analytical and numerical values is an artifact of the finite step size in the connection strength used for the simulation. If the simulations are repeated using a more strongly orientation-biased input ($\varepsilon = 0.3$), the contrast response curves do not change, whereas orientation tuning curves become continuously sharper with increasing connection strength. In either case, the separation of orientation tuning is contrast dependent outside the marginal phase.

We can summarize as follows: It can be shown very clearly, that a hypercolumn model with two neuron populations is not sufficient to account for the representation of stimulus contrast and stimulus orientation as observed in the primary visual cortex of many mammals. In the next section, we will introduce a mean-field model, which accounts for the large variety of different inhibitory neuron types observed in the cortex of cats and monkeys (Bartsch et al., 1999a; Bartsch et al., 1999c; Bartsch et al., 2000a).

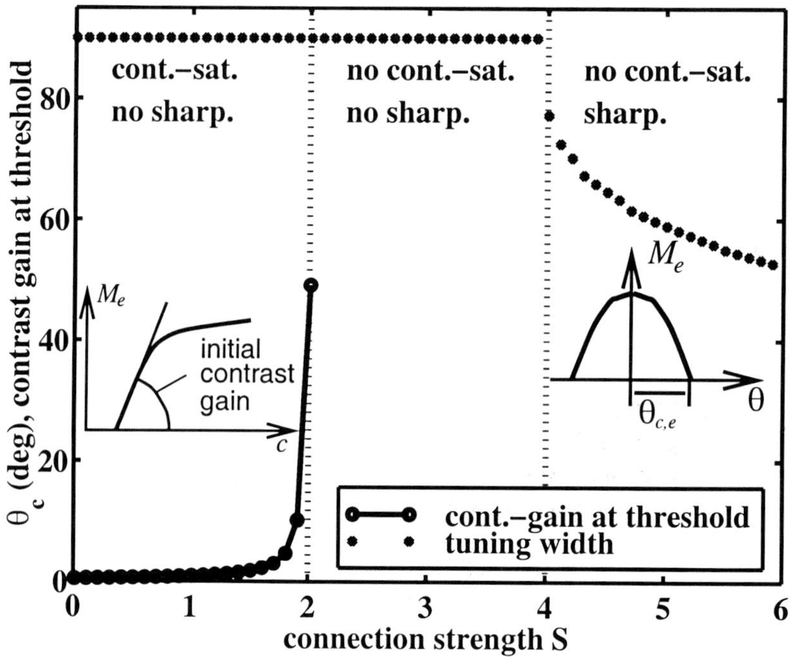


*Figure 10.10.* The behavior of the contrast gain at activation threshold (solid line) and the orientation tuning width (crosses) as a function of the connection strength $S \equiv E_0 = E_2 = I_0$ ($I_2 = 0$) for a hypercolumn with cosine/flat connectivity. Vertical dotted lines mark the analytical results. Other parameters were: $\beta_e = 0.5, \beta_i = 1, T_e = 1, T_i = 2, \varepsilon = 0.01$, contrast for the orientation tuning width: $c = 2.0$. Insets illustrate criteria used for calculation of the curves. Both analytical and numerical results predict, that there is no overlapping regime of co-occurrence of contrast saturation and contrast-invariant orientation tuning.

## 5. Hypercolumns with Multiple Populations

Now we explore what is the possible influence of the diversity of neurons in the cortex on its functional characteristics. For this we extend the model of an orientation column, which is now provided with a more complex structure: We take into account the fact that cortical tissue contains many different cell types and combine each of these cell types to a separate population. In general, an orientation column, indexed by its preferred orientation $\theta$, now contains $N_e$ populations with different excitatory neuron types and $N_i$ different populations of inhibitory neurons (figure 10.11b). The $n$-th excitatory population is indexed by $(e, n)$ and the $n$-th inhibitory population by $(e, n)$. We henceforth refer to the subpopulations as model neurons or simply "neurons".

The strength of recurrent intracortical couplings is assumed to depend only on the source and target orientation columns but not on the particular target

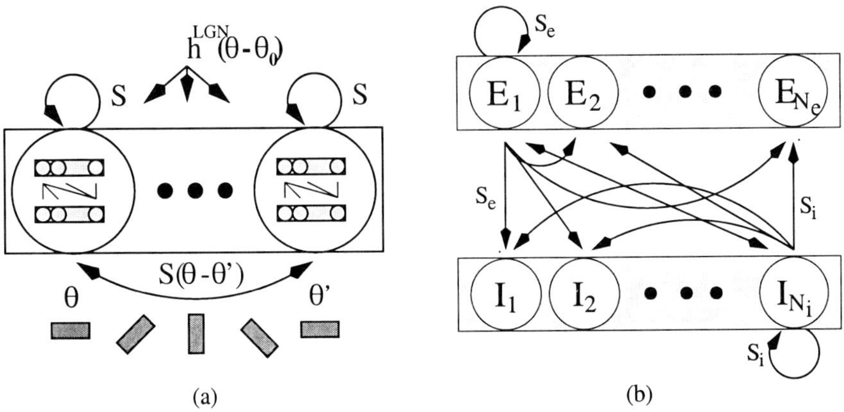


*Figure 10.11.* **(a)** Mean-field hypercolumn with many different cell types. Recurrent couplings depend on the source and target orientations only. **(b)** Structure of a single orientation column. It consists of $N_e$ excitatory and $N_i$ inhibitory cell populations with generally different properties.

neuron. The mean connection strength from neuron $(\alpha, n)$ within column $\theta'$ to neuron $\beta, m$ in column $\theta$ ($\alpha, \beta =$ e,i) is given by

$$S^{m,n}_{\beta,\alpha}(\theta, \theta') \equiv S_\alpha(\theta - \theta'). \tag{10.39}$$

The generalization compared to the previous section consists of the fact that different neuron subpopulations can have different mean cellular properties and wiring patterns. To start with a simple case, we keep all properties of the neurons up to their activation functions identical for the present considerations and assume that the neurons differ only in their mean activation thresholds. The activity of neuron $(\alpha, n)$, $\alpha = e, i$ in response to synaptic input $h$ is given by a semi-linear activation function

$$g_{\alpha,n}(h) = \max(\beta_\alpha(h - T_{\alpha,n}), 0), \tag{10.40}$$

where $\beta_\alpha$ denotes its slope and $T_{\alpha,n}$ its activation threshold. The activities of neurons $(e, n)$ and $(i, n)$ in column $\theta$, $m_{e,n}(\theta, t)$ and $m_{i,n}(\theta, t)$, evolve according to

$$\frac{d}{dt} m_{\alpha,n}(\theta, t) = -m_{\alpha,n}(\theta, t) + g_{\alpha,n}\left(h^{\mathrm{lat}}(\theta, t) + h^{\mathrm{LGN}}(\theta, t)\right) \tag{10.41}$$

$$h^{\mathrm{lat}}(\theta, t) = \sum_{\beta=e,i} \sum_{n} \int_{-\pi/2}^{\pi/2} d\theta' S_\alpha(\theta - \theta') m_{\beta,n}(\theta', t) \tag{10.42}$$

$$h^{\mathrm{LGN}}(\theta - \theta_0) = c(1 - \varepsilon + \varepsilon \cos(2(\theta - \theta_0))). \tag{10.43}$$

Note that $h^{\mathrm{LGN}}$ and $h^{\mathrm{lat}}$ are identical for all subpopulations.

## 5.1 Analytical Treatment of Contrast Saturation

We wish to understand how the contrast-response curve of the orientation column – or a representative subpopulation therein – depends on the distribution of activation thresholds. Again it seems reasonable to analyze an isolated but intrinsically coupled orientation column with $N_e$ excitatory and $N_i$ inhibitory neurons (figure 10.11b). In the stationary state, the total synaptic input, $H$, which is the same for all neurons in the orientation column, is given by

$$H = h^{\mathrm{LGN}} + S_e \sum_{n=1}^{N_e} M_{e,n} - S_i \sum_{n=1}^{N_i} M_{i,n}, \tag{10.44}$$

where, according to eq. (10.40), $M_{\alpha,n} = g_{\alpha,n}(H) \equiv M_{\alpha,n}(T_{\alpha,n}, H)$ are the steady state activations of the model neurons and $S_\alpha \equiv S_\alpha((\theta - \theta') = 0)$ abbreviate the identical intra-column connection strengths between the neurons. Now we assume, that the activation thresholds $T_e$ and $T_i$ are distributed over the orientation column according to pdfs $p_e(T_e)$ and $p_i(T_i)$, respectively. In the limit of infinitely many neurons, we can replace the sums in eq. (10.44) by the ensemble averages over the threshold distributions and obtain

$$\begin{aligned} H = h^{\mathrm{LGN}} \; &+ \; S_e \int_{-\infty}^{\infty} M_e(T_e, H) p_e(T_e)\, dT_e \\ &- \; S_i \int_{-\infty}^{\infty} M_i(T_i, H) p_i(T_i)\, dT_i. \end{aligned} \tag{10.45}$$

Because of the definition of the semi-linear transfer function eq. (10.40), we know that neurons with $T_\alpha \geq H$ are silent and therefore do not contribute to the sums or integrals in eqs. (10.44) and (10.45). Conversely, for $T_\alpha < H$, the activation function can be replaced by its linear part, $M_\alpha(T_\alpha, H) = \beta_\alpha(H - T_\alpha)$. Therefore we can replace the upper limits of the integrals in equation 10.45 by $H$:

$$H = h^{\mathrm{LGN}} + \beta_e S_e \int_{-\infty}^{H} (H - T_e) p_e(T_e)\, dT_e - \beta_i S_i \int_{-\infty}^{H} (H - T_i) p_i(T_i)\, dT_i. \tag{10.46}$$

Equation (10.46) represents a self-consistent relation between the total synaptic input $H$ and the afferent input $h^{\mathrm{LGN}}$. By solving this equation, we can write down an analytical solution for the stationary activation

$$M_e(T_e, H) \equiv M_e(T_e, H(h^{\mathrm{LGN}})) = M_e(T_e, h^{\mathrm{LGN}}) \tag{10.47}$$

as a function of the *external* instead of the total synaptic input, which is the contrast-response function of the neurons. Carrying out the integrals in eq.

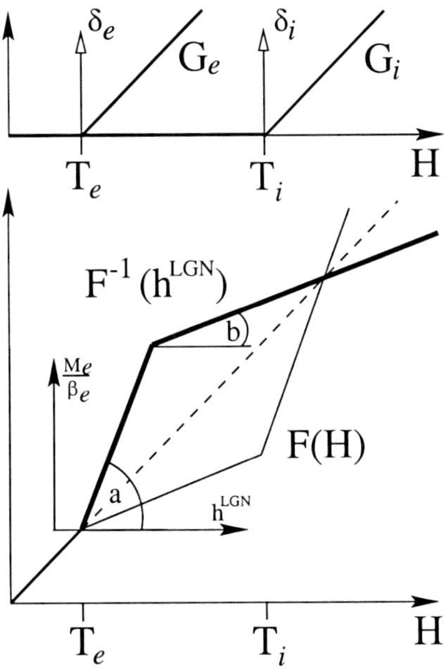


*Figure 10.12* Analytical solution eq. (10.51) for one isolated orientation column and $\delta$-peaked threshold distributions. **Top**: The distributions and the resulting second integrals $G_\alpha(H)$. **Bottom**: The function $F(H)$ eq. (10.50) (thin line) and its inverse (thick line) as resulting from the scenario in the top part. The thick line relative to the small coordinate system schematically illustrates the behavior of the contrast response function.

(10.46) yields

$$H = h^{\mathrm{LGN}} + S_e\beta_e G_e(H) - S_i\beta_i G_i(H) \tag{10.48}$$

$$G_\alpha(H) = \int_{-\infty}^{H} dH' \int_{-\infty}^{H'} dT p_\alpha(T), \quad \frac{d^2}{dT^2}G_\alpha(T) = p_\alpha(T), \tag{10.49}$$

$\alpha = e, i$. By defining the function

$$F(H) = H - \beta_e S_e G_e(H) + \beta_i S_i G_i(H) \tag{10.50}$$

equation 10.48 reduces to $F(H) = h^{\mathrm{LGN}}$ and we can express the steady state activations $M_\alpha$ by

$$M_\alpha(T_\alpha, h^{\mathrm{LGN}}) = \beta_\alpha (H - T_\alpha) = \beta_\alpha \left(F^{-1}\left(h^{\mathrm{LGN}}\right) - T_\alpha\right) \tag{10.51}$$

Equation (10.51) provides an analytical relationship between geniculate input and the response of the recurrent cortical circuit. Note that it only holds for one isolated orientation column and if $F$ is invertible. The latter condition corresponds to the boundary condition for the linear phase.

Figure 10.12 illustrates the meaning of eq. (10.51) for the special case of only one excitatory and one inhibitory neuron type and only two threshold values $T_e$ and $T_i$. In this case, the two threshold distributions reduce to Kronecker delta functions around the two thresholds, $p_e(T) = \delta(T - T_e)$ and $p_i(T) = \delta(T - T_i)$ and their second integrals become semilinear functions. $G_\alpha(H) = \max(H -$

$T_\alpha, 0)$ (figure 10.12 top). The function $F(H)$ (Eq. 10.50) becomes

$$F(H) = \begin{cases} H & H \le T_e \\ H - \beta_e S_e (H - T_e) & T_e < H \le T_i \\ H - \beta_e S_e (H - T_e) + \beta_i S_i (H - T_i) & H > T_i \end{cases} \quad (10.52)$$

In this scenario, the resulting contrast response function eq. (10.51) shows the typical saturating behavior, which we have already observed in the earlier treatment of contrast saturation (cf. section 4.1). The gradients of $F^{-1}(h^{\mathrm{LGN}})$ are $a = (1 - \beta_e S_e)^{-1}$, where $a\beta_e$ is the initial contrast gain of the contrast-response function, and $b = (1 - \beta_e S_e + \beta_i S_i)^{-1}$ for higher contrast levels. These expressions correspond to eqs. (10.30) and (10.34) for the special symmetries of the lateral connections considered here.

## 5.2 Numerical Simulations of Contrast Responses

For the following simulations we assumed threshold distributions, which for excitatory neurons are Gaussian, $p_e(T_e) = \mathcal{N}(\mu_e, \delta_e)$, and for inhibitory neurons are either also Gaussian, $p_i(T_i) = \mathcal{N}(\mu_i, \delta_i))$ or bimodal according to two superimposed Gaussian functions $p_i = 0.5(\mathcal{N}(\mu_{i,1}, \delta_{i,1}) + \mathcal{N}(\mu_{i,2}, \delta_{i,2}))$. Inhibitory mean activation thresholds are set to be higher ($\mu_i = 2$) than excitatory mean activation thresholds ($\mu_e = 1$). Also, simulations will use $\beta_e = 0.5$ and $\beta_i = 1$, but the special choice of parameters does not strongly influence the results.

Figure 10.13 compares the numerical solution of the differential equation eq. (10.41) (solid line) with the analytical expression eq. (10.51) (circles) for two unimodal and fairly narrow threshold distributions (histograms) in the linear phase. It demonstrates that the analytical solution approximates the solution of the differential equation very well. The dashed and dash-dotted lines plot $G_e$ and $G_i$ for the distributions used. The behavior of this system can be understood similarly to the contrast saturation treated in section 4.1: First, only excitatory neurons are active and, because we operate in the linear phase, act as linear amplifiers. For higher contrast levels, more and more inhibitors become active and reduce the contrast gain. Different from the case of only two thresholds, the contrast-response curve gradually changes its gain over contrast. A gradual contrast saturation can be qualitatively understood as follows: With increasing afferent input $h^{\mathrm{LGN}}$, more and more inhibitory neuron subpopulations are recruited (become active): The increase in number is proportional to $p_i(F^{-1}(h^{\mathrm{LGN}}))$. The more neurons are recruited, the stronger the decrease in contrast gain. In other words, we expect a relationship between the second derivative of the contrast-response function at $h^{\mathrm{LGN}}$ and the density of neurons with activation thresholds $T_i = F^{-1}(h^{\mathrm{LGN}})$.

This relationship can be quantified by by forming the 2nd derivative of the steady state activation eq. (10.51) with respect to the LGN input. We arrive

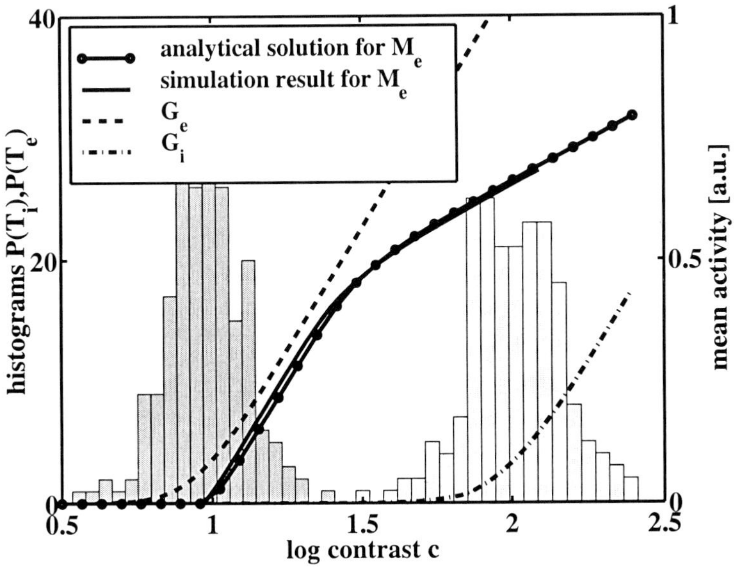


*Figure 10.13.* Simulation of a contrast-response curve for a set of 400 coupled model neurons (200 exc., 200 inh.) in the linear phase ($S_e = 1$, $S_i = 1$) and for unimodal Gaussian threshold distributions $p_e(T_e) = \mathcal{N}(1, 0.1)$, $p_i(T_i) = \mathcal{N}(2, 0.1)$ (cf. threshold histograms in the diagram). **Solid line**: Numerical solution of the differential equation eq. (10.41). **Circles**: Evaluation of the analytical expression (10.51). Both curves agree very well. Dashed and dash-dotted lines show $G_e$ and $G_i$, respectively.

at the following relationship between the curvature of the contrast-response function and the distributions of activation thresholds $p_\alpha(T_\alpha)$:

$$\frac{d^2}{d(h^{\rm LGN})^2} M_e = \beta_e \frac{S_e\beta_e p_e(H) - S_i\beta_i p_i(H)}{(-1 + S_e\beta_e G_e'(H) - S_i\beta_i G_i'(H))^3}, \quad (10.53)$$

$$H = F^{-1}(h^{\rm LGN}) \quad (10.54)$$

The denominator of eq. (10.53) is positive in the linear phase, because the gain of $F$ has to be finite (invertibility of $F$). The contrast-response curve shows a negative curvature or saturation, if more inhibitory than excitatory neurons are recruited by a small increase in the input, i.e. if $S_e\beta_e p_e(H) < S_i\beta_i p_i(H)$ holds. Otherwise, the contrast-response function increases its gain.

Besides a quantitative understanding of the structural origin of contrast gain in the linear phase, it might be even more important to see, whether a gradually increasing and finally saturating contrast-response can also be stabilized in the marginal phase by some threshold distribution. If this could be achieved, we would succeed in formulating necessary conditions for cortical circuitry to

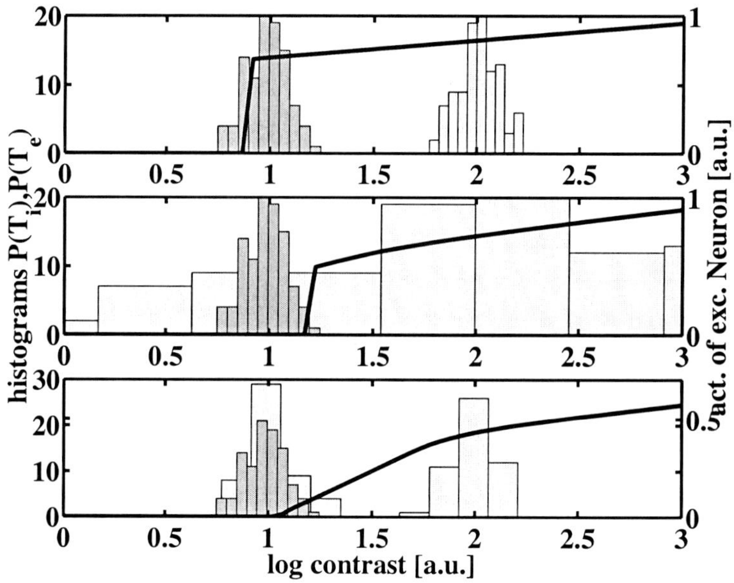


*Figure 10.14.* Simulation for a set of 400 coupled model neurons (200 exc., 200 inh.) in the marginal phase ($S_e = S_i = 6$). **Solid lines**: Contrast-response curves; **dark gray**: Histograms of excitatory thresholds $p_e(T_e) = \mathcal{N}(1, 0.1)$; **light gray**: Histograms of inhibitory thresholds $p_i(T_i)$. **Top**: $p_i(T_i)$ unimodal, small variance ($\mathcal{N}(2, 0.1)$); **Middle**: $p_i(T_i) = \mathcal{N}(2, 1)$ unimodal, large variance. **Bottom**: A bimodal distribution of $p_i(T_i)$ is used ($\mu_{i,1} = 1, \mu_{i,2} = 2, \delta_{i,1} = \delta_{i,2} = 0.1$). A bimodal distribution $p_i$ is necessary and sufficient for graded contrast-response and contrast saturation also in the marginal phase.

show a constant orientation tuning and contrast saturation for a single parameter setting.

Figure 10.14 shows the contrast-response curve of an excitatory neuron with threshold $T_e = 1$ for different cases of the inhibitory threshold distribution in the marginal phase. If the threshold-distribution is small and unimodal (top), the contrast-response shows a pseudo-binary switch-on behavior as observed in the marginal phase with two neuron types (cf. section 4.3). This behavior remains stable, as long as the distribution is unimodal, even if it is very wide (figure 10.14, middle). As soon as the threshold distribution becomes bimodal (figure 10.14, bottom), the contrast-response first increases from zero and later saturates, as observed in biology. This demonstrates that two inhibitory neuron populations, one with low and the other with higher activation threshold, are necessary and sufficient to stabilize contrast saturation in the marginal phase.

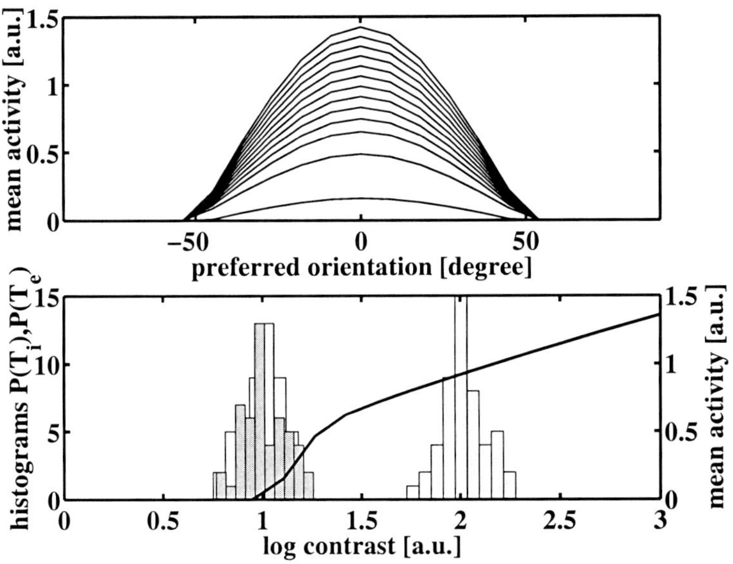


*Figure 10.15.* **Top** Orientation tuning curve and **Bottom** contrast-response curve of an excitatory neuron with preferred orientation $0^o$ for a hypercolumn with 21 orientation columns (50 excitatory neurons and 100 inhibitory neurons each) in the marginal phase. The system shows a graded and saturating contrast response, which is combined with a contrast-invariant orientation tuning width. Parameters: $S_e = 6, S_i = -6, \sigma_e = 34$ deg, $\sigma_i = \infty$, $\delta_e = \delta_{i,1} = \delta_{i,2} = 0.1, \mu_e = \mu_{i,1} = 1, \mu_{i,2} = 2$.

## 5.3 Orientation and Contrast Response with Three Neuron Types

Now we are ready to combine many structured orientation columns as analyzed previously to a full hypercolumn. The orientation columns are mutually coupled by lateral connections with Gaussian profiles as specified in eq. (10.13), and are driven by weakly orientation biased input ($\varepsilon = 0.1$).

Figure 10.15 shows orientation tuning curves of the $m_{e,1}(\theta)$ excitatory populations of 21 orientation columns (top) and the contrast-response curves of a subset of 5 excitatory subpopulations of the $\theta = 0^o$ column (bottom) for unimodal $p_e(T_e)$ and bimodal $p_i(T_i)$. Even though the system operates in the marginal phase, where orientation tuning is independent of contrast, the contrast-response curve shows expressed saturation at the same time. This behavior is independent of the detailed shape of the threshold distributions, as long as it is bimodal.

A phase-diagram determined by the initial contrast gain and the orientation sharpening (cf. figure 10.10) for a hypercolumn with one excitatory and two inhibitory (low- and high-threshold) neuron types is plotted in Figure 10.16.

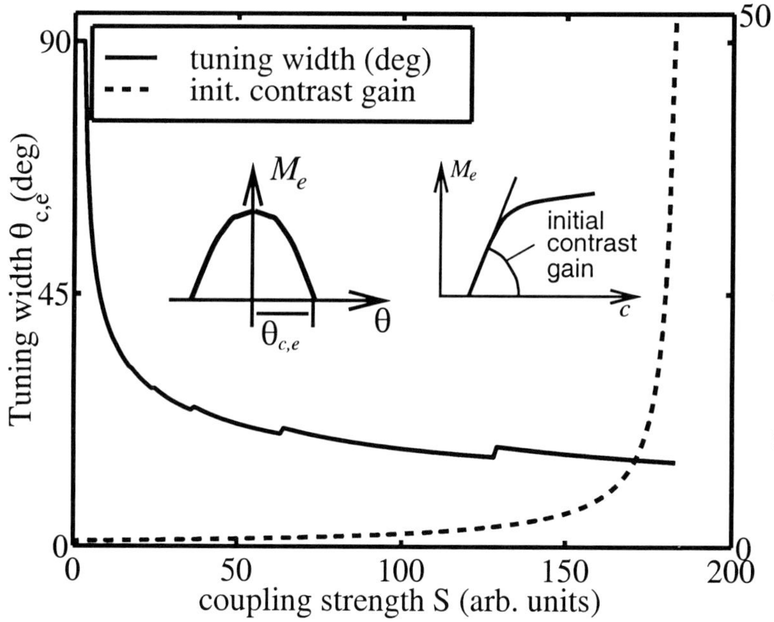


*Figure 10.16.* The behavior of the contrast gain at activation threshold (solid line) and the orientation tuning width (crosses) as a function of the connection strength $S \equiv E_0 = E_2 = I_0$ ($I_2 = 0$) for a hypercolumn with one excitatory and two inhibitory neuron populations. Other parameters were: $\beta_e = 0.5, \beta_i = 1, T_e = 1, T_{i1} = 1, T_{i2} = 2, \varepsilon = 0.01$, contrast for the orientation tuning width: $c = 2.0$. Insets illustrate criteria used for calculation of the curves. There is a wide range ($4 \leq S \leq 180$), over which the linear and marginal phase coincide. Steps in the solid line are finite-size effect.

Due to low-threshold inhibition, the linear phase with finite initial contrast gain is stabilized up to very strong recurrent excitation strengths ($S \approx 180$ compared to $S = 2$ for two-neuron hypercolumns), and there is a wide range of coupling strengths, in which orientation tuning is invariant and the contrast-response saturates. In summary, this finding predicts that the experimentally observed cortical response properties require essentially two functionally distinct inhibitory neuron types to be present: Inhibitors with a low activation threshold (or tonically active inhibitors) stabilize the contrast gain at or near the contrast threshold to finite values, whereas inhibitors with high activation thresholds cause the saturation of the contrast-response curves at higher contrast levels.

In figure 10.15, both types of inhibitors were assumed to distribute lateral inhibition between different orientations. Possible candidates for such inhibitors are basket cells with axonal arborizations up to 1200 $\mu$m (Lund, 1987), but

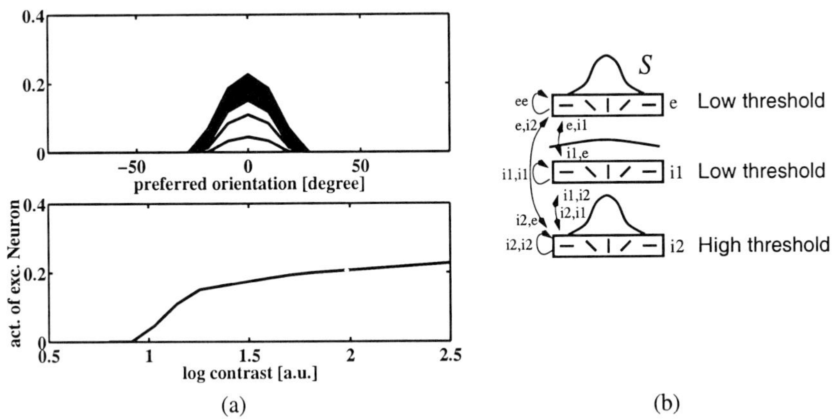


*Figure 10.17.* **(a)** Orientation tuning curves (top) and the contrast-response function of the zero-deg. orientation column (bottom). **(b)** Schematic illustration of the corresponding wiring scheme: low-threshold lateral inhibitors (e.g. basket neurons) and high-threshold local inhibitors (e.g. chandelier cells). Parameters: $S_e = S_{i1} = S_{i2} = 50$; (marginal phase) **(b)** $\sigma_e = 34$ deg; $(\sigma_{i1}, \sigma_{i2}) = (\infty, 34)$ deg. $T_e = T_{i1} = 1; T_{i2} = 1.5$. The hypercolumn properly operates as in figure 10.15.

it seems more reasonable to identify the two functionally different inhibitors with two anatomically distinguishable biological neuron types. Many inhibitors apart from basket cells are local companions, which contact only postsynaptic neurons within the same or closely adjacent orientation columns. One important local inhibitor is the chandelier cell. Therefore, we may ask, under which conditions a hypercolumn with pyramidal neurons as the excitatory population, basket cells as lateral inhibitors and chandelier cells as local inhibitors, still show the behavior seen in figure 10.15.

Orientation tuning and contrast response for a hypercolumn with three neuron types are provided in figures 10.17 and 10.18 for two different combinations of wiring profiles and activation thresholds of inhibitory neurons. If the low-threshold cells mediate lateral inhibition and the high-threshold neurons local inhibition (figure 10.17b), orientation tuning is sharp and constant with saturating contrast response function (figure 10.17a). If the properties are reversed (low-threshold chandelier cells and high-threshold basket cells, figure 10.18b), orientation sharpening is weak, unstable and contrast-dependent (figure 10.18a). These simulations emphasize the important role of inhibition for the observed cortical representation of orientation and contrast of a stimulus (cf. (Eysel et al., 1998)), but additionally provide the following prediction: A hypercolumn needs two different inhibitors for the generation of experimentally observed contrast and orientation representation. At least one of the cell types must mediate lateral inhibition (e.g. basket cells), and this cell type must have a

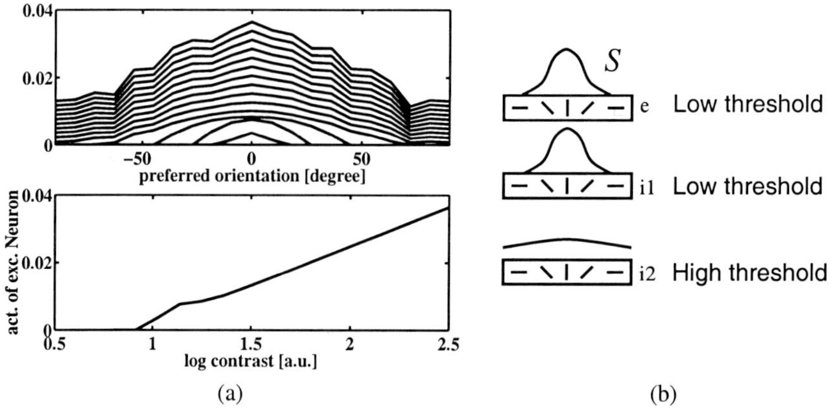


*Figure 10.18.* **(a)** Orientation tuning curves (top) and the contrast-response function of the zero-deg. orientation column (bottom) for reverse properties of the inhibitory neurons: **(b)** low-threshold local inhibitors and high-threshold lateral inhibitors. Parameters: $S_e = S_{i1} = S_{i2} =$ 50; (marginal phase) **(b)** $\sigma_e = 34$ deg; $(\sigma_{i1}, \sigma_{i2}) = (34, infty)$ deg. $T_e = T_{i1} = 1; T_{i2} = 1.5$.

low activation threshold. If local inhibitors (e.g. chandelier cells) contribute to the recurrent circuit as modeled, they should have a high activation threshold.

## 6. Mean-Field Modeling of Contextual Effects

As summarized in chapter 3, section 2.3, oriented stimuli outside the classical receptive field modulate the responses to an oriented stimulus within the classical receptive field of a considered cell without being able to drive the cell alone. Showing an iso-oriented annular surround stimulus outside the classical receptive field of a neuron additionally to an centered stimulus suppresses its response (Blakemore and Tobin, 1972; Sillito et al., 1995; Levitt and Lund, 1997). This behavior might be a possible mechanism for texture-based segmentation, where contour is defined by an abrupt change in the orientation of an elongated texture. Figure 10.19a shows an example for a visual scene, in which the boundary between two extended gratings with different orientations pops out. One possible mechanism for the amplification would be an increased response of orientation-selective neurons with receptive fields near the border. Those neurons would see different orientations within and outside their classical receptive fields, and would have an increased response. In contrast, neurons far from the border would see the same orientation within and outside their classical receptive field and their responses would be decreased by iso-orientation suppression.

But in a different stimulus paradigm, namely if the non-classical receptive field is stimulated by small flanking grating patches or bars instead of full an-

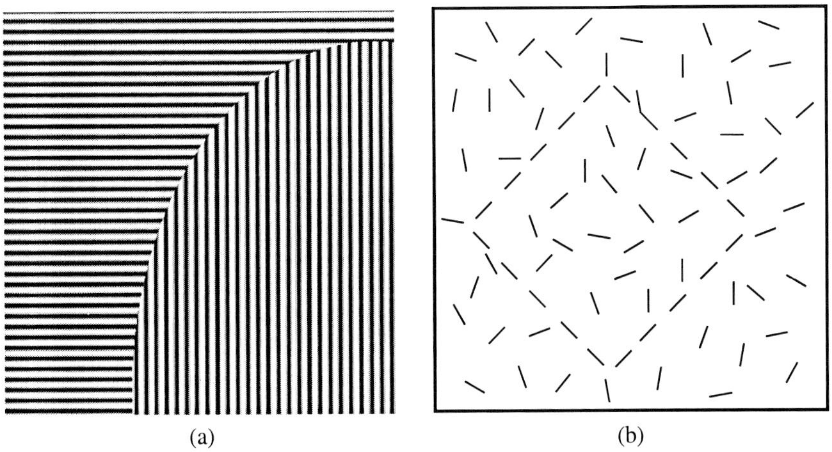


*Figure 10.19.* **(a)** Example image for the demonstration of texture-based segmentation (a contour is defined by texture boundaries). **(b)** Example image for the demonstration of line-completion (aligned line segments are perceptually grouped to an interrupted diamond).

nuli, iso-oriented surround stimuli can also facilitate the response of a neuron (Sengpiel et al., 1997; Polat et al., 1998; Kapadia et al., 1999). This observation, which apparently contradicts the previous findings, could serve as the physiological basis of line completion, which is schematically illustrated in figure 10.19b. Line segments which are aligned are perceptually linked together to parts of a continuous contour, and we perceive an interrupted circle.

One possible anatomical substrate mediating this interactions are orientation specific long-range connections formed by excitatory pyramidal neurons (figure 3.9) (Rockland and Lund, 1983). Models incorporating these patchy connections showed that nonclassical receptive field effects can be mediated by these fibers (Pawelzik et al., 1996; Mundel et al., 1996; Todorov et al., 1996; Bartsch et al., 1997; Stetter et al., 2000a; Bartsch et al., 2000b). In this section we summarize some approaches towards an understanding of the role of these connections for the formation of contextual effects in V1.

## 6.1 Model Setup for Contextual Effects

One approach towards a model of contextual effects consists of two neighboring and coupled mean-field hypercolumns $a = 1, 2$ within the primary visual cortex (but see (Bartsch et al., 2000c) for an alternative approach). Hypercolumn 1 is considered to process the visual input within the considered receptive field and is referred to as "center" hypercolumn. The aggregate field of hypercolumn 2 is assumed to be adjacent but still disjunct from the considered receptive field. It processes the nonclassical receptive field of the "center" hy-

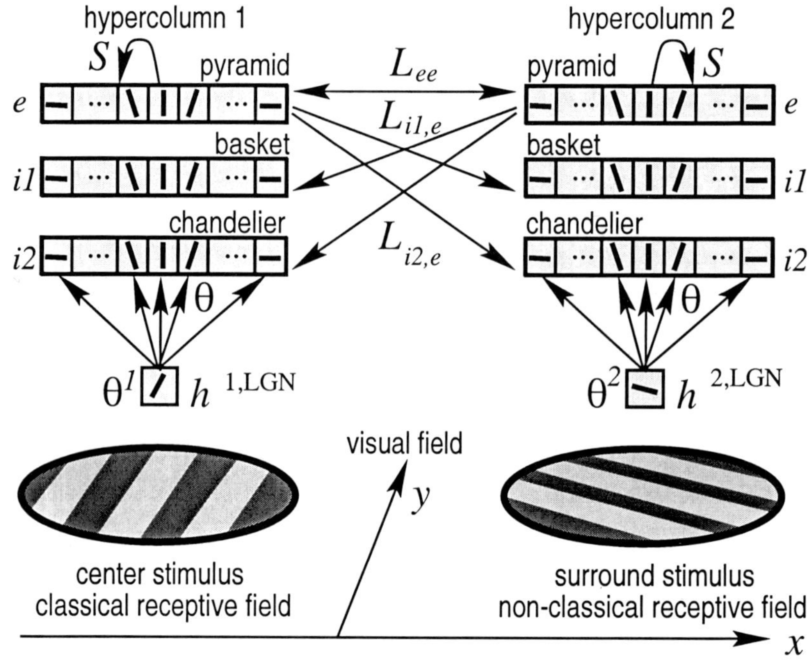


*Figure 10.20.* Mean-field model of two coupled hypercolumns $a = 1, 2$, the orientation columns $\theta$ of which contain one excitatory ('e') and two inhibitory ('i1','i2') neuron populations. Both hypercolumns receive weakly orientation biased geniculocortical inputs $h^{a,\mathrm{LGN}}$, $a = 1, 2$, from adjacent but nonoverlapping patches of the visual scene, which correspond to the center and the nonclassical surround of hypercolumn 1. Orientation columns within each hypercolumn are densely interconnected by short range connections $S_{\alpha,\beta}(\theta - \theta')$, where $\alpha$ denotes the type of the target population and $\beta$ the type of the source population ($\alpha, \beta =$ 'e', 'i'). In addition, both hypercolumns are mutually interconnected by symmetrical and excitatory long-range connections $L_{\alpha,\beta}(\theta - \theta')$.

percolumn and modulates it via their mutual couplings. Figure 10.20 schematically illustrates the model setup.

Again, each hypercolumn consists of a set of orientation columns, indexed by their preferred orientations $\theta$, and each column consists of an excitatory $(e)$ and two inhibitory neuron populations $(i1, i2)$. The activity of neuron $(\alpha)$, $\alpha = e, i1, i2$ in response to synaptic input $h$ is given by a semi-linear activation function $g_\alpha(h) = \max(\beta_\alpha(h - T_\alpha), 0)$ where $\beta_\alpha$ denotes its slope and $T_\alpha$ its activation threshold. Similarly to eq. (10.16), the dynamics of a neuron population $\alpha$ in hypercolumn $a$ and column $\theta$, $m_e^a(\theta, t)$, $m_{i1}^a(\theta, t)$ and $m_{i2}^a(\theta, t)$, are described by the following set of differential equations:

$$\frac{d}{dt} m_\alpha^a(\theta, t) \quad = \quad -m_\alpha^a(\theta, t) + g_\alpha\left(h^{a,\mathrm{lat}}(\theta, t) + h^{a,\mathrm{LGN}}(\theta, t)\right) \quad (10.55)$$

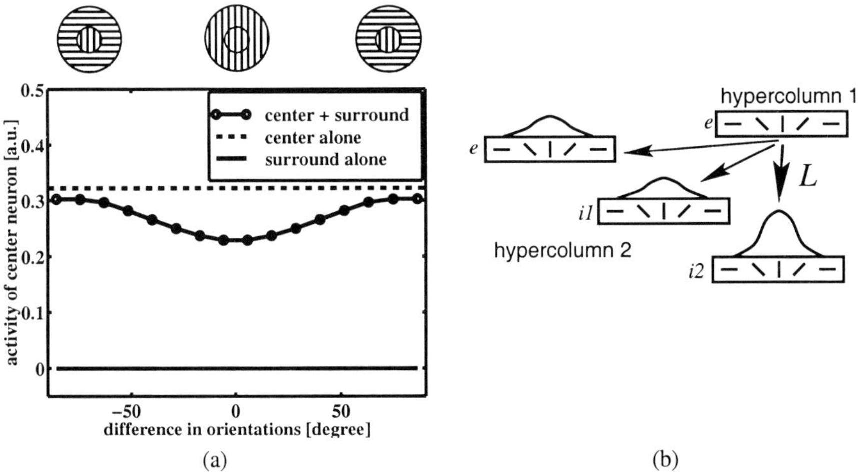


*Figure 10.21.* **(a)** Modulation of the center response by an oriented stimulus in the nonclassical surround. Compared to stimulation of the center hypercolumn alone (dashed line), the surround stimulus causes iso-orientation suppression (circles), but has only a weak impact in the cross-orientation stimulus condition. The surround stimulus alone cannot drive (solid line but only modulate the center hypercolumn. **(b)** Connectivity needed for the behavior in (a). Long-range connections must predominantly drive inhibitory interneurons for iso-orientation suppression. Parameters were $L_{e;i1;i2} = 0.5, 0.5, 3$, $\lambda_{\alpha\beta} \equiv \lambda_\beta = 34$ deg, $\beta = e, i1, i2$.

$$
\begin{aligned}
h^{a,\mathrm{lat}}(\theta, t) &= \sum_{\beta=e,i1,i2} \int_{-\pi/2}^{\pi/2} d\theta' \Big[ S_{\alpha,\beta}(\theta-\theta') m^a_\beta(\theta', t) + \\
&\qquad L_{\alpha,\beta}(\theta-\theta') m^{b\neq a}_\beta(\theta', t) \Big] \qquad (10.56)\\
h^{a,\mathrm{LGN}}(\theta) &= c(1 - \varepsilon + \varepsilon \cos(2(\theta - \theta^a))), \qquad (10.57)
\end{aligned}
$$

where $\theta^a$ is the stimulus orientation presented to the center hypercolumn ($a = 1$) or to the surround hypercolumn ($a = 2$). Intracortical couplings are symmetric between both hypercolumns and all long-range connections are excitatory, i.e. $L_{\alpha,\beta} = L_{\alpha,e} =: L_\alpha$. They depend only on the difference in preferred orientations and are assumed as Gaussian functions in orientation space:

$$
\begin{aligned}
S_{\alpha,\beta}(\Delta\theta) &= \mathrm{sign}_\beta S_{\alpha\beta} N_\sigma \exp\left( -\frac{(\Phi(\Delta\theta))^2}{2\sigma^2_{\alpha\beta}} \right) \qquad (10.58)\\
L_\alpha(\Delta\theta) &= L_\alpha N_\lambda \exp\left( -\frac{(\Phi(\Delta\theta))^2}{2\lambda^2_\alpha} \right), \qquad (10.59)
\end{aligned}
$$

where $L_\alpha \geq 0$ is the integral strength of the long-range connections to population $\alpha$ and $N_\lambda$ is a normalization constant.

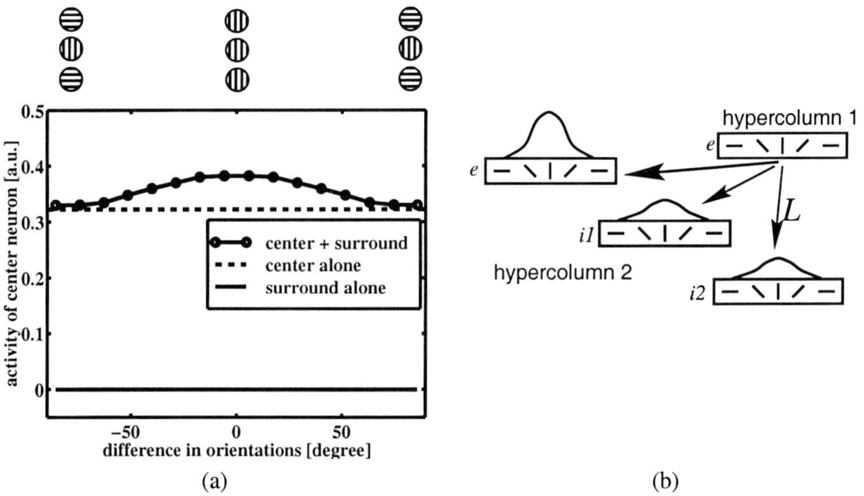


*Figure 10.22.* **(a)** Iso-orientation facilitation for the same circuit as in figure 10.21, but this time the long-range connections drive excitatory target neurons stronger than the inhibitors **(b)**. Parameters: $L_{e;i1;i2} = 1, 0.5, 0.5$, $\lambda_\beta = 34$ deg, $\beta = e, i1, i2$.

## 6.2 Numerical Simulations

Based on the coupled hypercolumn model we can now explore, if and how long-range connections can modulate local cortical processing. For the following simulations, we used a strong local recurrent connectivity with identical strengths $S_\alpha = 50$ and widths $\sigma_{\alpha\beta} \equiv \sigma_\beta$, $\sigma_e = \sigma_{i2} = 34$ deg, $\sigma_{i1} = \infty$ (cf. figure 10.17). Afferent input had intermediate orientation bias $\varepsilon = 0.3$ and $c = 2.5$, and the parameters for the activation functions were chosen as $\beta_e = 0.5$, $\beta_{i1} = \beta_{i2} = 1$, $T_e = T_{i1} = 1$ and $T_{i2} = 1.5$. The results reported do not qualitatively depend on these choices as long as the system operates in the overlap region of the linear and marginal regimes (central part in figure 10.16).

Figure 10.21a demonstrates, how a stimulus presented in the non-classical surround of hypercolumn 1 (the center hypercolumn) can modulate its response to a stimulus within the receptive field. Compared to the dashed line, which marks its response to center stimulation alone, the activity of the center column is reduced (circles), if an oriented stimulus is presented to the nonclassical surround (see icons above plot). If center and surround orientations are identical or similar, the suppression is strongest, i.e. this system shows iso-orientation suppression. In contrast, if both stimuli are orthogonal to each other, only a weak suppressive effect is observed. In particular, over all orientation differences, the sign of the modulatory effect is the same. The solid line in 10.21a shows the response of the center hypercolumn to surround stimulation alone and demonstrates, that the surround stimulus cannot activate but only modulate the neurons

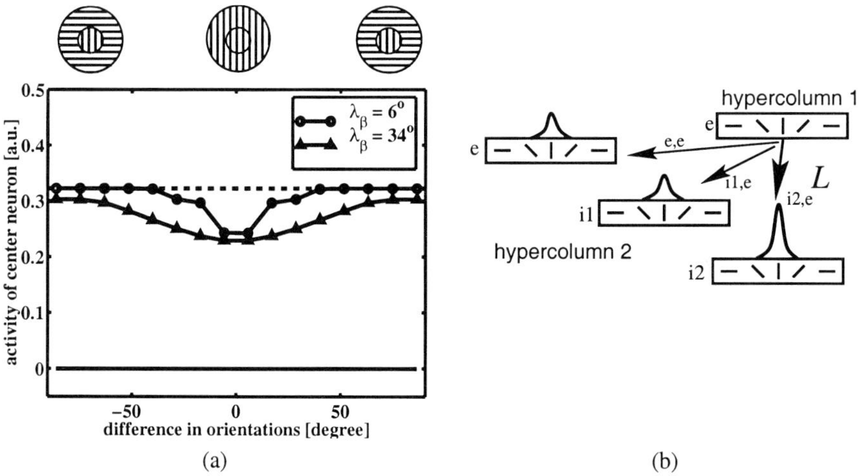


*Figure 10.23.* **(a)** Dependence of the suppression profile on the orientation-specificity of long-range connections. Circles: Strongly orientation specific long-range couplings ($\lambda_\beta = 6$ deg) cause a narrowly tuned iso-orientation suppression. Triangles show the same curve as in figure 10.21 ($\lambda_\beta = 6$ deg) for comparison. Again, the dashed and solid lines mark the response to center alone and surround alone stimulation. **(b)** Schematic illustration of the long-range connectivity used. Parameters: $L_{e;i1;i2} = 0.5, 0.5, 3$, $\lambda_\beta = 6$ deg, $\beta = e, i1, i2$.

in the center hypercolumn. Modulatory suppression as shown in figure 10.21a requires a particular connection scheme for the long-range connections, which is summarized in figure 10.21b: $(i)$ Long-range connections should predominantly connect columns with similar preferred orientations, which is supported by experiments (Malach et al., 1993; Bosking et al., 1997). $(ii)$ The fibers must drive at least one inhibitory neuron type stronger than the excitatory populations which has been suggested by recent experiments (Das and Gilbert, 1999). For high contrast levels, where all neuron populations are active, the effect does not depend on which inhibitory neuron type (i1 or i2) is driven strongest.

Figure 10.22 shows a parameter regime, in which the surround stimulus facilitates the center response (circles vs. dashed line), but cannot drive the neurons of the center hypercolumn alone (solid line). Again, cross-orientation modulation is weak and has the same sign as the center modulation. Facilitation is observed, if the long-range connections drive excitatory target neurons stronger than inhibitory ones (figure 10.22b).

The angular profile of the nonclassical modulation is mostly determined by the orientation specificity of the long-range couplings $L_{\alpha,\beta}(\theta - \theta')$. This is demonstrated in figure 10.23a, which compares the suppressive modulation caused by strongly orientation specific long-range connections (circles) with suppression for more broadly tuned long-range connectivity (triangles). The

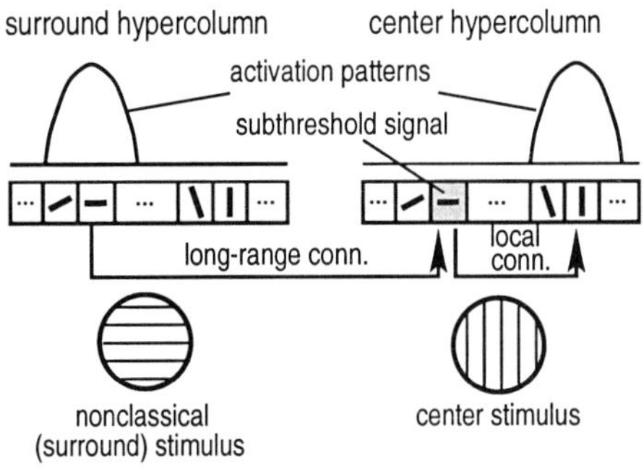


*Figure 10.24* Cross-orientation modulation in the marginal phase. Iso-orientation specific long-range connections cannot evoke any cross-orientation modulation, because all basket cells that signal across orientations are silenced by the local recurrent circuitry (shaded orientation column).

orientation specificity of long-range connections is strongly correlated with the orientation tuning of the suppression. In contrast, the profile of non-classical modulation depends only weakly on the local connectivity within the hypercolumns, as long as they operate in the marginal phase (data not shown).

In order to understand, why the modulatory effects are determined predominantly by the properties of the long-range connections, we have to realize that the activity pattern within each hypercolumn is determined by the local recurrent circuitry: Because the hypercolumns operate in the marginal regime, the local circuit forms a sharply tuned activity patch around the orientation column which matches the stimulus orientation (figure 10.24). The curve shape of the patch is relatively rigid and can only be weakly influenced by external (afferent or lateral) input. In particular, any synaptic input that is mediated by long-range connections can only modulate the activity level of active neurons, but cannot activate silent neurons. This behavior is schematically illustrated in figure 10.24 for cross-orientation stimulation. The shaded orientation column is driven by long-range connections, but cannot become active because its state is determined by the local recurrent dynamics. Consequently, we can only expect a non-classical modulation to occur, if there are long-range fibers which connect active source neurons with active target neurons, in other words, the range of a surround modulation in orientation space is approximately given by the width $\lambda$ of the long-range connection profile plus the width of the cortical activity pattern. Because the activity patterns have approximately constant shape, the angular profile of the surround modulation is determined by the angular profile of the long-range connections in orientation space.

Figure 10.24 also helps to understand, why cross-orientation modulation is hard to achieve with iso-orientation specific patchy connections. It sketches the situation of cross-oriented stimuli and purely iso-orientation specific long-range fibers. The activity patterns of source and target neurons are disjunct in orientation space, and therefore no cross-orientation effect can be observed. In particular, orientation contrast sensitivity (iso-orientation suppression com-

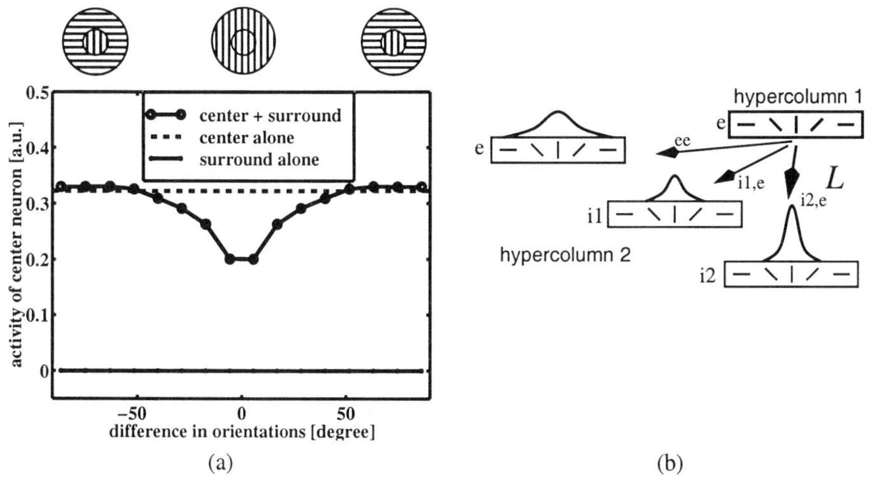


*Figure 10.25.* **(a)** Iso-orientation suppression combined with a weak cross-orientation facilitation appears, as soon as long-range connections to excitatory target neurons are more broadly tuned than connections to inhibitory target neurons **(b)**. Parameters: $L_{e;i1;i2} = 0.5, 0.5, 3$, $\lambda_{i1} = \lambda_{i2} = 17$ deg, $\lambda_e = 34$ deg.

bined with cross-orientation facilitation) in the marginal phase cannot be caused by dis-inhibition as suggested earlier (Pawelzik et al., 1996).

Figure 10.25a shows a simulation, in which iso-orientation suppression is combined with a weak cross-orientation facilitation (sensitivity to orientation-contrast). This behavior is caused by long-range connections, which are more broadly tuned for excitatory target neurons than for inhibitory target neurons (figure 10.25b). As a consequence, long-range modulation via inhibitory interneurons dominates at small orientation differences between source and target orientation column, whereas for larger differences in orientation direct excitation dominates. In other words, the profile of the long-range connections implements an inverse Mexican hat in orientation space, which directly translates into orientation-contrast sensitivity.

## 7. Concluding Remarks

Because individual hypercolumns within the model show a very different behavior depending on the parameter regime (either linear or marginal), it is important to determine, which phase, if any, might be implemented in the primary visual cortex. Unfortunately, this cannot be tested directly, because the phase boundaries provided above cannot be translated directly into biologically accessible quantities, and also because the total strength between biological neuron populations cannot be easily measured. However, in both the linear and the marginal phase there are model predictions which can help testing whether biological brain states can be properly described by a mean-field model in one or the other regime. Both regimes have some advantages and drawbacks.

In the marginal phase, salient features (e.g. high contrast oriented gratings) are amplified and represented with a high signal-to-noise ratio, and the stimulus quality is decoupled from the stimulus strength. Less salient features (e.g low-contrast oriented gratings) are suppressed, which may not always be desirable. A further property of a hypercolumn which acts in the marginal phase is amplification even of small random fluctuations: Even untuned input would cause a sharp orientation-selective activity pattern, as soon as it exceeds the threshold. Because neural transmission is noisy, blobs in the marginal phase would be present all the time and change their position on the cortical surface in response to orientation biased input rather than being switched on and off due to input.

Conversely, if the cortex would operate in the linear phase, the strength of cortical activation would be sensitive to changes in the afferent activities and would establish a useful internal representation, which represents also the details of a visual scene. Simulations have shown, that a hypercolumn acting in the linear phase (as opposed to the marginal phase) seems better capable of representing more than one stimulus orientation simultaneously and showing cross-stimulation effects (Stetter et al., 2000a). Recently, fast synaptic depression of intracortical connections has been proposed as a mechanism to combine the advantages of both phases. Immediately after a saccade, the synapses are strong, the cortex operates in the marginal phase and reliably extracts the most salient features from the highly noisy input, which consists only of few spikes at that time. Later, as more and more afferent spikes are accumulated, the synapses become weaker due to fast depression, the system approaches the linear phase and represents the details of the visual scene (Adorján et al., 2000; Schwabe et al., 2000).

One possible mechanism for contrast saturation is the presence of high threshold inhibition (as observed in (McCormick et al., 1985)). At a given contrast level, inhibitory neurons with a high activation threshold are recruited, inhibit the postsynaptic excitatory neurons and cause their saturation. If this mechanism is indeed realized in cortex, it would be accompanied by the following set of testable effects: $(i)$ There are inhibitory neurons, which become active only for high contrast levels at which pyramidal neurons start to saturate. $(ii)$ These inhibitory neurons themselves do not saturate but show roughly linear and steep contrast response functions up to 100 % contrast.

Cross-orientation contextual effects may occur $(i)$ if there exist direct long-range connections between neurons with orthogonal preferred orientations (this is not experimentally supported so far) or $(ii)$ if long-range connections are iso-orientation specific, but contact local interneurons with low-thresholds, which in turn project to orthogonal orientations and thus provide the link between both

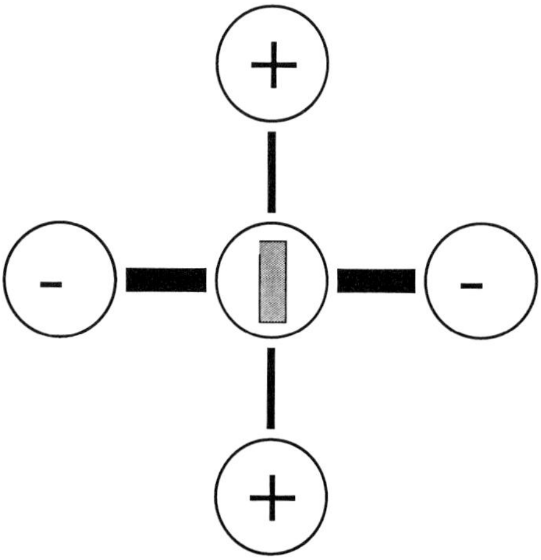


*Figure 10.26* Proposed wiring pattern: Long-range collaterals drive pyramidal neurons along the retinotopic direction which corresponds to the preferred orientation of the source neuron, and strongly drive inhibitory interneurons in the orthogonal direction.

populations. These hypothetical interneurons must not participate in the tight local cortical feedback within one hypercolumn, since otherwise they would be silenced by the inhibitory input provided by the active orthogonal populations. In effect, the long-range couplings should provide super-threshold input to these interneurons which in turn would have modulatory effects to the populations with orthogonal preferred orientations. Thus, if the latter scenario would be realized in the cortex, there should be excitatory neurons with low activation thresholds, which are directly activated (not only modulated) by long-range connections.

The exploration of the neuronal implementation of contextual effects by a hypercolumn-based mean-field model was also able to resolve the apparent contradiction between $(i)$ experiments which report iso-orientation suppression by annular surround stimuli and support the hypothesis that V1 contributes to texture-based segmentation, and $(ii)$ studies which report iso-orientation facilitation by co-aligned flanking patch stimuli and support the hypothesis that V1 contributes to line completion. Figure 10.26 displays a proposed wiring pattern for cortical patchy connections, which can explain both phenomena in the same anatomical substrate (Li, 1998; Bartsch et al., 2000b). If a long oriented bar is presented as stimulus, the area it covers can be projected from the visual field to the cortical surface via their retinotopic relationship – it defines a visuotopic axis on the cortical surface. It is proposed that long-range fibers that emerge from a neuron weakly drive excitatory target neurons along the visuotopic axis of its own preferred orientation and strongly drive inhibitory interneurons along the orthogonal orientation.

If the stimulus in the non-classical surround covers a whole annulus around the classical receptive field, both the collinear facilitatory and the strong orthogonal suppressive connections are recruited. In this scenario, the whole

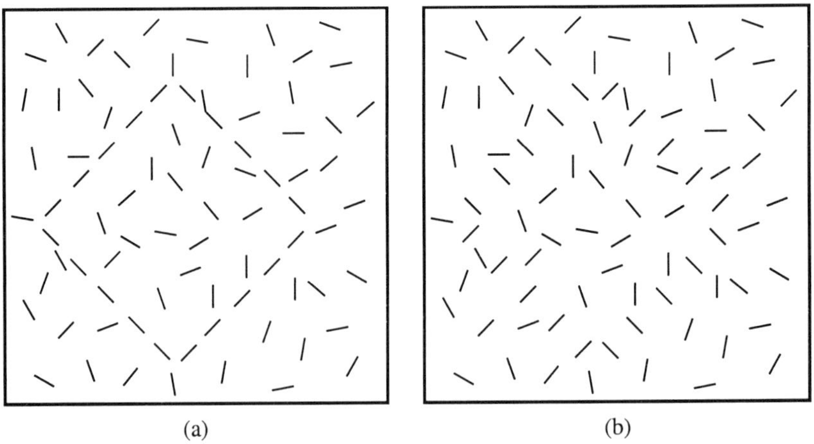

*Figure 10.27.* Perceptual grouping of line segments requires a collinear arrangement. **(a)** If line segments are aligned, they are preceptually grouped to an interrupted diamond (same as in figure 10.19b). **(b)** Same arrangement as in a, but the co-aligned lines are rotated by 90 deg.The rotated lines are not grouped.

cortical tissue which represents the annular area of the surround stimulus is identified with the surround hypercolumn. This is possible because the hypercolumn model does not include any statement about the spatial distribution of orientation columns that contribute to a hypercolumn. For this interpretation of the surround hypercolumn, the long-range interactions on average drive inhibitory interneurons more strongly than excitatory target neurons, and the model correctly predicts iso-orientation suppression. Conversely, if the non-classical surround stimulus consists of two flanking bars which are co-aligned both visuotopically and in orientation with the center stimulus, the surround hypercolumn must be identified with the two patches of cortex that represent the two flanking stimuli only. In this case, only the collinear facilitative long-range connections are recruited, and the hypercolumn model correctly predicts iso-orientation facilitation.

At least one paradigm of perceptual grouping seems compatible with (and might be causally related to) the proposed wiring scheme of figure 10.26 and is demonstrated by the two line assemblies in figure 10.27. All lines in both squares are identical, except the lines which belong to the diamond. These segments are co-aligned in figure 10.27a and rotated by 90 deg (but still aligned) in figure 10.27b. The co-aligned segments are strongly grouped, but grouping has almost completely disappeared in the scene with the rotated line segments. Perceptual grouping is strongly anisotropic, and is strongest if the segments are co-aligned. Line completion also does not require attentive search or any concious effort. It is a pre-attentive phenomenon of human perception, which indicates that early vision might play an important role in its generation. One possible neuronal correlate for perceptual grouping could be manifested in a lateral wiring scheme of the primary visual cortex as proposed in figure 10.26. Future investigations will answer the questions, whether this scheme is implemented in early vision, if and to what extent it is part of the general construction principles of the neocortex, and what is the general design principle it implements.

# References


Abbott, L. F. and Kepler, T. B. (1990). *Model Neurons: From Hodgkin-Huxley to Hopfield*, pages 6–18. Springer, Berlin.

Adorján, P., Schwabe, L., Piepenbrock, C., and Obermayer, K. (2000). Recurrent cortical competition: Strengthen or weake? In Solla, S. A., Leen, T. K., and Müller, K.-R., editors, *Advances in Neural Information Processing Systems*, volume 12, page in press. MIT Press.

Ahmed, B., Anderson, J. C., Douglas, R. J., Martin, K. A., and Whitteridge, D. (1998). Estimates of the net excitatory currents evoked by visual stimulation of identified neurons in cat visual cortex. *Cereb. Cortex*, 8:462–476.

Albrecht, D. G. (1995). Visual cortex neurons in monkey and cat: Effect of contrast on the spatial and temporal phase transfer function. *J. Neurosci.*, 12:1191–1210.

Albrecht, D. G. and Hamilton, D. B. (1982). Striate cortex of monkey and cat: contrast response function. *J. Neurophysiol.*, 48:217–237.

Almeida, R. and Stetter, M. (2001). Relating functional imaging and neuronal population spiking activity. *Soc. Neurosci. Abstr.*, 27:in press.

Amari, S. (1996). Neural learning in structured parameter spaces - natural riemannian gradient. In Mozer, M. C., Jordan, M. I., and Petsche, T., editors, *Advances in Neural Information Processing Systems*, volume 9.

Andrew, R. D. and MacVicar, B. A. (1994). Imaging cell volume changes and neuronal excitation in the hippocampal slice. *Neuroscience*, 62:371–383.

Artola, A., Bröcher, S., and Singer, W. (1990). Different voltage-dependent thresholds for inducing long-term depression and long-term potentiation in slices of rat visual cortex. *Nature*, 347:69–72.

Atick, J. J. (1992). Could information theory provide an ecological theory of sensory processing. *Network*, 3:213–251.

Atick, J. J., Li, Z., and Redlich, A. N. (1990). Color coding and its interaction with spatiotemporal processing in the retina. Technical report, Inst. of Advanced Study.

Atick, J. J. and Redlich, A. N. (1990). Towards a theory of early visual processing. *Neural Comput.*, 2:308–320.

Azouz, R., Gray, C. M., Nowak, L. G., and McCormick, D. A. (1997). Physiological properties of inhibitory interneurons in cat striate cortex. *Cereb. Cortex*, 7:534–545.

Babel, D. B., Timberlake, G. T., Schuchard, R. A., Maino, J. H., and Stetter, M. (1997). SLO magnification changes due to ammetropia can be corrected optically. In *Invest. Ophthalm. Vis. Sci.*

Bartsch, H., Stetter, M., and Obermayer, K. (1997). A model for orientation tuning and contextual effects of orientation selective receptive fields. In Gerstner, W., Germond, A., Hasler, M., and Nicoud, J.-D., editors, *Artificial Neural Networks - ICANN '97*, volume 1327 of *Lecture notes in computer science*, pages 237–242. Springer Berlin.

Bartsch, H., Stetter, M., and Obermayer, K. (1999a). About the influence of neuronal variability in a mean-field model of the visual cortex. In Elsner, N. and Eysel, U., editors, *From molecular neurobiology to clinical neuroscience.* Thieme Stuttgart.

Bartsch, H., Stetter, M., and Obermayer, K. (1999b). Mean-field modeling of cortical contrast saturation and orientation tuning. *Soc. Neurosci Abstr.*, 25:1933.

Bartsch, H., Stetter, M., and Obermayer, K. (1999c). On the influence of threshold variability in a model of the visual cortex. In *Artificial Neural Networks – ICANN '99*, pages 73–78.

Bartsch, H., Stetter, M., and Obermayer, K. (2000a). The influence of threshold variability on the response of visual cortical neurons. *Neurocomputing*, page in press.

Bartsch, H., Stetter, M., and Obermayer, K. (2000b). Modeling contextual effects in macaque V1 by recurrent networks with multiple inhibitors. In *Proceedings of the CNS*2000 conference, Brugge*, page in press.

Bartsch, H., Stetter, M., Weber, C., and Obermayer, K. (2000c). Influence of the geometry of lateral connections in V1 on orientation selectivity: A model study. *Soc. Neurosci. Abstr.*, page submitted.

Bauer, C., Burger, T., Stetter, M., and Obermayer, K. (2000). A neural network model for the self-organization of cortical grating cells. *Z. Naturforsch.*, page in press.

Bauer, R., Eckhorn, R., and Jordan, W. (1989). Iso- and cross-oriented columns in cat striate cortex: a study with simultaneous single- and multi-unit recordings. *Neuroscience*, 30:733–740.

Bauer, U. (1999). *Computational models of neural circuitry in the macaque monkey primary visual cortex.* Cuvillier Göttingen.

Bell, A. J. and Sejnowski, T. J. (1995). An information-maximization approach to blind separation and blind deconvolution. *Neural Comput.*, 7:1129–1159.

Ben-Yishai, R., Bar-Or, R. L., and Sompolinsky, H. (1995). Theory of orientation tuning in visual cortex. *Proc. Natl. Acad. Sci. USA*, 92:3844–3848.

Ben-Yishai, R., Hansel, D., and Sompolinsky, H. (1997). Traveling waves and the processing of weakly tuned inputs in a cortical network module. *J. Comput. Neurosci.*, 4:57–77.

Bi, G. Q. and Poo, M. M. (1998). Synaptic modifications in cultured hippocampal neurons: dependence on spike timing, synaptic strength, and postsynaptic cell type. *J. Neurosci.*, 18:10464–10472.

Bishop, C. M. (1995). Neural networks for pattern recognition. Clarendon Press, Oxford.

Blakemore, C. and Tobin, E. A. (1972). Lateral inhibition between orientation detectors in the cat's visual cortex. *Exp. Brain Res.*, 15:439–440.

Blasdel, G. G. (1992a). Differential imaging of ocular dominance and orientation selectivity in monkey striate cortex. *J. Neurosci.*, 12:3115–3138.

Blasdel, G. G. (1992b). Orientation selectiviy, preference, and continuity in monkey striate cortex. *J. Neurosci.*, 12:3139–3161.

Blasdel, G. G. and Fitzpatrick, D. (1984). Physiological organization of layer 4 in macaque striate cortex. *J. Neurosci.*, 4:880–895.

Blasdel, G. G. and Lund, J. S. (1983). Termination of afferent axons in macaque striate cortex. *J. Neurosci.*, 3:1389–1413.

Blasdel, G. G. and Salama, G. (1986). Voltage-sensitive dyes reveal a modular organization in monkey striate cortex. *Nature*, 321:579–585.

Bliss, T. V. P. and Collingridge, G. L. (1993). A synaptic model of memory: long-term potentiation in the hippocampus. *Nature*, 361:31–39.

Bonhoeffer, T. and Grinvald, A. (1991). Iso-orientation domains in cat visual cortex are arranged in pinwheel-like patterns. *Nature*, 353:429–431.

Bonhoeffer, T. and Grinvald, A. (1993). The layout of iso-orientation domains in area 18 of cat visual cortex: optical imaging reveals a pinwheel-like organization. *J. Neurosci.*, 13:4157–4180.

Bonhoeffer, T. and Grinvald, A. (1996). Optical imaging based on intrinsic signals: The methodology. In Toga, A. and Maziotta, J. C., editors, *Brain mapping: The methods*, pages 55–97, San Diego, CA. Academic Press, Inc.

Bonhoeffer, T., Kim, D. S., Malonek, D., Shoham, D., and Grinvald, A. (1995). Optical imaging of the layout of functional domains in area 17 and across the area 17/18 border in cat visual cortex. *Eur. J. Neurosci.*, 7:1973–1988.

Bonhoeffer, T., Staiger, V., and Aertsen, A. (1989). Synaptic plasticity in rat hippocampal slice cultures: Local "hebbian" conjunction of pre- and postsynaptic stimulation leads to distributed synaptic enhancement. *Proc. Natl. Acad. Sci. USA*, 86:8113–8117.

Bosking, W. H., Zhang, Y., Schofield, B., and Fitzpatrick, D. (1997). Orientation selectivity and the arrangement of horizontal connections in tree shrew striate cortex. *J. Neurosci.*, 17:2112–2127.

Bower, J. M. and Beeman, D. (1994). *The book of GENESIS*. Springer, New York.

Bressloff, P. C. (1994). Dynamics of compartmental model recurrent neural networks. *Phys. Rev. E*, 50(3):2308–2319.

Bressloff, P. C. and Taylor, J. G. (1993a). Compartmental-model response function for dendritic trees. *Biol. Cybern.*, 70:199–207.

Bressloff, P. C. and Taylor, J. G. (1993b). Spatiotemporal pattern processing in a compartmental model neuron. *Phys. Rev. E*, 47(4):2899–2912.

Brown, T. H., Kairiss, E. W., and Keenan, C. L. (1990). Hebbian synapses: Biophysical mechanisms and algorithms. *Annu. Rev. Neurosci.*, 13:475–511.

Brunner, K., Kussinger, M., Stetter, M., and Lang, E. W. (1997). A neural network model for cortical grating cells. In Elsner, N. and Wässle, H., editors, *From membrane to mind.* Thieme Stuttgart.

Brunner, K., Kussinger, M., Stetter, M., and Lang, E. W. (1998). A neural network model for the emergence of grating cells. *Biol. Cybern.*, 78:389–397.

Bucher, D., Scholz, M., Stetter, M., Obermayer, K., and Pflüger, H.-J. (2000). Correction methods for three-dimensional reconstructions from confo cal images. *J. Neurosci. Meth.*, 100:135–143.

Buchsbaum, G. and Gottschalk, A. (1983). Trichromacy, opponent colours coding and optimum colour information transmission in the retina. *Proc. R. Soc. Lond. B*, 220:89–113.

Bullier, J. and Henry, G. H. (1980). Ordinal position and afferent input of neurons in monkey striate cortex. *J. Comp. Neurol.*, 193:913–935.

Burton, G. J. and Moorhead, I. R. (1987). Color and spatial structure in natural scenes. *Appl. Opt.*, 26:157–170.

Bygrave, F. L. and Benedetti, A. (1996). What is the concentration of calcium ions in the endoplasmatic reticulum? *Cell Calcium*, 19:547–551.

Callaway, E. M. (1998). Local circuits in primary visual cortex of the macaque monkey. *Annu. Rev. Neurosci.*, 21:47–74.

Cannestra, A. F., Blood, A. J., Black, K. L., and Toga, A. W. (1996). The evolution of optical signals in human and rodent cortex. *NeuroImage*, 3:202–208.

Cardoso, J. F. (1997). Infomax and maximum likelihood for blind source separation. *IEEE Signal Processing Lett.*

Cardoso, J. F. and Souloumiac, A. (1996). Numerical methods for simultaneous diagonalization. *SIAM J. Mat. Anal. Appl.*, 17:161.

Carmona, R. A., Hwang, W. L., and Frostig, R. D. (1995). Wavelet analysis for brain function imaging. *IEEE Trans. Med. Imaging*, 14:556–564.

Chapman, B., Stryker, M. P., and Bonhoeffer, T. (1996). Development of orientation preference maps in ferret primary visual cortex. *J. Neurosci.*, 16:6443–6453.

Cheong, W.-F., Prahl, S. A., and Welch, A. J. (1990). A review of the optical properties of biological tissues. *IEEE J. Quant. Electron.*, 26:2166–2185.

Coenen, A. M. L. and Vendrik, A. J. H. (1972). Determination of the transfer ratio of cat's geniculate neurons through quasi-intracellular recordings and the relation with the level of alertness. *Exp. Brain Res.*, 14:227–242.

Cohen, L. B. (1973). Changes in neuron structure during action potential propagation and synaptic transmission. *Physiol. Review*, 53:373–418.

Cohen, L. B., Keynes, R. D., and Hille, B. (1968). Light scattering and birefringence changes during nerve activity. *Nature*, 218:438–441.

Cohen, M. S. and Bookheimer, S. Y. (1994). Localization of brain function using magnetic resonance imaging. *Trends Neurosci.*, 17:268–277.

Comon, P. (1994). Independent component analysis: A new concept? *Signal Processing*, 36:287–314.

Connolly, M. and Van Essen, D. (1984). The representation of the visual field in parvocellular and magnocellular layers of the lateral geniculate nucleus in the macaque monkey. *J. Comp. Neurol.*, 226:544–564.

Crick, F. (1984). Function of the thalamic reticular complex: The searchlight hypothesis. *Proc. Natl. Acad. Sci. USA*, 81:4586–4590.

Croner, L. J. and Kaplan, E. (1995). Receptive fields of P and M ganglion cells across the primate retina. *Vision Res.*, 35:7–24.

Dalva, M. B., Weliky, M., and Katz, L. C. (1997). Relationships between local synaptic connections and orientation domains in primary visual cortex. *Neuron*, 19:871–880.

Damasio, A. R. (1996). *Descartes' error.* Papermac MacMillan London.

Das, A. and Gilbert, C. D. (1995). Long-range horizontal connections and their role in cortical reorganization revealed by optical recording of cat primary visual cortex. *Nature*, 375:780–784.

Das, A. and Gilbert, C. D. (1999). Topography of contextual modulations mediated by short-range interactions in primary visual cortex. *Nature*, 399:655–661.

Daugman, J. G. (1980). Two-dimensional spectral analysis of cortical receptive field profiles. *Vision Res.*, 20:847–856.

Daw, N. W., Stein, P. S. G., and Fox, K. (1993). The role of nmda receptors in information processing. *Annu. Rev. Neurosci.*, 16:207–222.

DeAngelis, G. C., Robson, J. G., Ohzawa, I., and Freeman, R. D. (1992). Organization of suppression in receptive fields of neurons in cat visual cortex. *J. Neurophysiol.*, 68:144–163.

Deco, G. and Obradovic, D. (1996). *An information-theoretic approach to neural computing.* Springer, New York.

Deco, G. and Schürmann, B. (2000). *Information dynamics: Foundations and Applications.* Springer, New York.

DeMonasterio, F. M. and Gouras, P. (1975). Functional properties of ganglion cells of the rhesus monkey retina. *J. Physiol.*, 251:167–195.

Deriugin, N. G. (1956). The power spectrum and the correlation function of the television signal. *Telecommunications*, 1:1–12.

Destexhe, A., Contreras, D., Steriade, M., Sejnowski, T. J., and Huguenard, J. R. (1996). In vivo, in vitro, and computational analysis of dendritic calcium currents in thalamic reticular neurons. *J. Neurosci.*, 161:169–185.

Dinse, H. R., Reuter, G., Cords, S. M., Godde, B., Hilger, T., and Lenarz, T. (1997). Optical imaging of cat auditory cortical organization after electrical stimulation of a multichannel cochlear implant: differential effects of acute and chronic stimulation. *Am. J. Otol.*, 18:S17–S18.

Dow, B. M. (1974). Functional classes of cells and their laminar distribution in monkey visual cortex. *J. Neurophysiol.*, 37:927–946.

Dowling, J. E. (1992). *Neurons and networks. An introduction to neuroscience.* Harvard University Press, Cambridge.

Eckhorn, R., Bauer, R., Jordan, W., Brosch, M., Kruse, W., Munk, M., and Reitboeck, H. J. (1988). Coherent oscillations: a mechanism of feature linking in the visual cortex? multiple electrode and correlation analyses in the cat. *Biol. Cybern.*, 60:121–130.

Eckhorn, R. and Pöpel, B. (1974). Rigorous and extended application of information theory to the afferent visual system of the cat i. basic concepts. *Kybernetik*, 16:191–200.

Eckhorn, R. and Pöpel, B. (1975). Rigorous and extended application of information theory to the afferent visual system of the cat ii. experimental results. *Biol. Cybern.*, 17:7–17.

Edwards, D. P., Purpura, K. P., and Kaplan, E. (1995). Contrast sensitivity and spatial frequency response of primate cortical neurons in and around the cytochrome oxidase blobs. *Vision Res.*, 35:1501–1523.

Engert, F. and Bonhoeffer, T. (1997). Synapse specifity of long-term potentiation breaks down at short distances. *Nature*, 388:279–284.

Ermentrout, B. (1996). Type i membranes, phase resetting curves, and synchrony. *Neural Computation*, 8:979–1001.

Ermentrout, G. B. and Kopell, N. (1998). Fine structure of neural spiking and synchronization in the presence of conduction delays. *Proc. Natl. Acad. Sci. USA*, 95:1259–1264.

Everson, R., Knight, B. W., and Sirovich, L. (1997). Separating spatially distributed response to stimulation from background. i. optical imaging. *Biol. Cybern.*, 77:407–417.

Everson, R. M., Prashanth, A. K., Gabbay, M., Knight, B. W., Sirovich, L., and Kaplan, E. (1998). Representation of spatial frequency and orientation in the visual cortex. *Proc. Natl. Acad. Sci.*, 95:8334–8338.

Eysel, U. T., Shevelev, I. A., Lazareva, N. A., and Sharaev, G. A. (1998). Orientation tuning and receptive field structure in cat striate neurons during local blockade of intracortical inhibition. *Neuroscience*, 84:25–36.

Field, D. J. (1987). Relations between the statistics of natural images and the response properties of cortical cells. *J. Opt. Soc. Am. A*, 4(12):2379–2394.

Field, D. J. (1994). What is the goal of sensory coding. *Neural Comput.*, 6:559–601.

Fitzhugh, R. (1960). Thresholds and plateaus in the hodgkin-huxley nerve equations. *J. General Physiol.*, 43:867–896.

Fitzhugh, R. (1961). Impulses and physiological states in theoretical models of nerve membrane. *Biophys. J.*, 1:445–466.

Fitzke, F. W. and Masters, B. R. (1993). Three-dimensional visualization of confocal sections of in vivo human fundus and optic nerve. *Current Eye Research*, 12:1015–1018.

Fitzke, F. W., Woon, H., Timberlake, G., Robinson, L., Marshall, J., and Bird, A. C. (1991). Optical modifications to a scanning laser ophthalmoscope for high magnification, narrow optical section imaging. *Lasers and Light in Ophthalmology*, 4:7–14.

Flock, S. T., Patterson, M. S., Wilson, B. C., and Wyman, D. R. (1989a). Monte carlo modeling of light propagation in highly scattering tissues-i: Predictions and comparison with diffusion theory. *IEEE Trans. Biomed. Eng.*, 36:1163–1168.

Flock, S. T., Wilson, B. C., and Patterson, M. S. (1989b). Monte carlo modeling of light propagation in highly scattering tissues-ii: Comparison with measurements in phantoms. *IEEE Trans. Biomed. Eng.*, 36:1169–1173.

Friel, D. D. and Tsien, R. W. (1992). A caffeine- and ryanodine-sensitive Ca2+ store in bullfrog sympathetic neurones modulates effects of Ca2+ on [Ca2+]i. *J. Physiol.*, 450:217–246.

Fromherz, P. and Müller, C. O. (1994). Cable properties of a straight neurite of a leech neuron probed by a voltage sensitive dye. *Proc. Natl. Acad. Sci. USA*, 91:4604–4608.

Fromherz, P. and Vetter, T. (1992). Cable properties of arborized retzius cells of the leech in culture as probed by a voltage-sensitive dye. *Proc. Natl. Acad. Sci. USA*, 89:2041–2045.

Fyfe, C. and Baddeley, R. (1995). Finding compact and sparse-distributed representations of visual images. *Network*, 6:333–344.

Galizia, C. G., Joerges, J., Küttner, A., Faber, T., and Menzel, R. (1997). A semi-in-vivo preparation for optical recording of the insect brain. *J. Neurosci. Methods*, 76:61–69.

Galizia, C. G., Nägler, K., Hölldobler, B., and Menzel, R. (1998). Odour coding is bilaterally symmetrical in the antennal lobes of honeybees (apis mellifera). *Eur. J. Neurosci.*, 10:2964–2974.

Galizia, C. G., Sachse, S., Rappert, A., and Menzel, R. (1999). The glomerular code for odor representation is species specific in the honeybee Apis mellifera. *Nature Neurosci.*, 2:473–478.

Gardiner, C. W. (1983). *Handbook of stochastic methods*. Springer Berlin Heidelberg New York Tokyo.

Gilbert, C. D. and Wiesel, T. N. (1989). Columnar specifity of intrinsic horizontal and cortico-cortical connections in cat visual cortex. *J. Neurosci.*, 9:2432–2442.

Gilbert, C. D. and Wiesel, T. N. (1990). The influence of contextual stimuli on the orientation selectivity of cells in primary visual cortex of the cat. *Vision Res.*, 30:1689–1701.

Godde, B., Spengler, F., and Dinse, H. R. (1996). Associative pairing of tactile stimulation induces somatosensory cortical reorganization in rats and humans. *Neuroreport*, 8:281–285.

Gödecke, I. and Bonhoeffer, T. (1996). Development of identical orientation maps for two eyes without common visual experience. *Nature*, 379:251–254.

Gödecke, I., Kim, D.-S., Bonhoeffer, T., and Singer, W. (1997). Development of orientation preference maps in area 18 of kitten visual cortex. *Eur. J. Neurosci.*, 9:1754–1762.

Graaf, R., Dassel, A. C. M., Koelink, M. H., Aarnoudse, J. G., de Mul, F. F. M., Zijlstra, W. G., and Greve, J. (1993a). Condensed monte carlo simulations applied to reflectance pulse oximetry. In Chance, B. and Alvano, R. R., editors, *Photon Migration and Imaging in Random Media and Tissues.*, volume 1888, pages 201–213. SPIE.

Graaf, R., Koelink, M. H., de Mul, F. F. M., Zijlstra, W. G., Dassel, A. C. M., and Arnoudse, J. G. (1993b). Condensed monte carlo simulations for the description of light transport. *Appl. Optics*, 32:426–435.

Gray, C. M., König, P., Engel, A. K., and Singer, W. (1989). Oscillatory responses in cat visual cortex exhibit inter-columnar synchronization which reflects global stimulus properties. *Nature*, 338:334–337.

Greve, H., Stetter, M., Galizia, C. G., Menzel, R., and Obermayer, K. (1999a). Model-based analysis of $ca^{2+}$ imaging data from the antennal lobe of the honeybee. In Elsner, N. and Eysel, U., editors, *From molecular neurobiology to clinical Neuroscience*, page 365. Thieme, Stuttgart 1999.

Greve, H., Stetter, M., Galizia, C. G., and Obermayer, K. (1999b). Nonlinear models applied to the analysis of Ca-dynamics in the antennal lobes of honeybees. *Soc. Neurosci. Abstr.*, 25:1313.

Grinvald, A., Lieke, E., Frostig, R. D., Gilbert, C. D., and Wiesel, T. N. (1986). Functional architecture of cortex revealed by optical imaging of intrinsic signals. *Nature*, 324:361–364.

Guillery, R. W., Feig, S. L., and Lozsádi, L. A. (1998). Paying attention to the thalamic reticular nucleus. *TINS*, 21:28–32.

Hammond, P. and Andrews, D. P. (1978). Orientation tuning of cells in areas 17 and 18 of the cat's visual cortex. *Exp. Brain Res.*, 31:341–351.

Hansel, D. and Sompolinsky, H. (1998). Modeling feature selectivity in local cortical circuits. In Koch, C. and Segev, I., editors, *Methods in Neural Modeling*, chapter 13, pages 499–567. MIT Press, Cambridge MA.

Hawken, M. J., Parker, A. J., and Lund, J. S. (1988). Laminar orgnaization and contrast sensitivity of direction-selective cells in the striate cortex of the old world monkey. *J. Neurosci.*, 8:3541–3548.

Hebb, D. O. (1949). *The organization of behavior*. Wiley, New York.

Henyey, L. G. and Greenstein, J. L. (1941). Diffuse radiation in the galaxy. *Astrophys. J.*, 93:70–83.

Hess, A. and Scheich, H. (1996). Optical and fdg mapping of frequency-specific activity in auditory cortex. *Neuroreport*, 7:2643–2647.

Hodgkin, A. L. (1948). The local electric changes associated with repetitive action in a non-medullated axon. *J. Physiol.*, 107:165–181.

Hodgkin, A. L. and Huxley, A. F. (1952). A quantitative description of membrane current and its application to conduction and excitation of nerve. *J. Physiol.*, 117:500–544.

Horton, J. C. (1984). Cytochrome oxidase patches: A new cytoarchitectonic feature of monkey visual cortex. *Phil. Trans. R. Soc. Lond.*, B304:199–253.

Hubel, D. H. and Wiesel, T. N. (1962). Receptive fields, binocular interaction and functional architecture in the cat's visual cortex. *J. Physiol.*, 160:106–154.

Hubel, D. H. and Wiesel, T. N. (1968). Receptive fields and functional architecture of monkey striate cortex. *J. Physiol.*, 195:215–243.

Hubel, D. H. and Wiesel, T. N. (1974). Uniformity of monkey striate cortex: A parallel relationship between field size, scatter, and magnification factor. *J. Comp. Neur.*, 158:295–306.

Hubel, D. H. and Wiesel, T. N. (1977). Functional architecture of macaque monkey visual cortex. *Proc. R. Soc. Lond. B*, 198:1–59.

Hübener, M., Shoham, D., Grinvald, A., and Bonhoeffer, T. (1997). Spatial relationships among three columnar systems in cat area 17. *J. Neurosci.*, 17:9270–9284.

Hyvärinen, A., Karhunen, J., and Oja, E. (2001). *Independent Component Analysis*. John Wiley & Sons New York.

Hyvärinen, A. and Oja, E. (1997). A fast fixed point algorithm for independent component analysis. *Neural Comput.*, 9:1483–1492.

Jack, J. J. B., Noble, D., and Tsien, R. W. (1975). *Electric current flow in excitable cells*. Clarendon Press, Oxford.

Joerges, J., Küttner, A., Galizia, C. G., and Menzel, R. (1997). Representation of odours and odour mixtures visualized in the honeybee brain. *Nature*, 387:285–288.

Johnston, D., Magee, J. C., Colbert, C. M., and Christie, B. R. (1996). Active properties of neural dendrites. *Annu. Rev. Neurosci.*, 19:165–186.

Jones, J. P. and Palmer, L. A. (1987a). An evaluation of the two-dimensional gabor filter model of simple receptive fields in cat striate cortex. *J. Neurophysiol.*, 58:1233–1258.

Jones, J. P. and Palmer, L. A. (1987b). The two-dimensional spatial structure of simple receptive fields in cat striate cortex. *J. Neurophysiol.*, 58:1187–1211.

Jones, J. P., Stepnoski, A., and Palmer, L. A. (1987). The two-dimensional spectral structure of simple receptive fields in cat striate cortex. *J. Neurophysiol.*, 58:1212–1232.

Jutten, C. and Herault, J. (1986). *Neural Network for Computing*. AIP, New York.

Kandel, E. R., Schwartz, J. H., and Jessel, T. M. (1991). *Principles of Neural Sciences*. Prentice Hall International London.

Kaneko, A. (1979). Physiology of the retina. *Annu. Rev. Neurosci.*, 2:169–191.

Kapadia, M. K., Ito, M., Gilbert, C. D., and Westheimer, G. (1995). Improvement in visual sensitivity by changes in local context: parallel studies in human observers and in V1 of alert monkeys. *Neuron*, 15:843–856.

Kapadia, M. K., Sigman, M., and Gilbert, C. D. (1999). Lateral interactions in cortical area V1 and their role in perception. *Soc. Neurosci. Abstr.*, 25(1):1049.

Kaplan, E. and Shapley, R. M. (1982). X and Y cells in the lateral geniculate nucleus of macaque monkeys. *J. Physiol. Lond.*, 330:125–143.

Kay, S. M. (1993). *Fundamentals of statistical signal processing. Estimation theory*. Prentice Hall International Editions, New Jersey.

Kim, D. S. and Bonhoeffer, T. (1994). Reverse occlusion leads to a precise restoration of orientation preference maps in visual cortex. *Nature*, 370:370–372.

Kim, D.-S., Duong, T. Q., and Kim, S.-G. (2000a). High-resolution mapping of iso-orientation columns by fMRI. *Nature Neurosci.*, 3:164–169.

Kim, D.-S., Duong, T. Q., and Kim, S.-G. (2000b). Reply to "can current fMRI techniques reveal the micro-architecture of the cortex?". *Nature Neurosci.*, 3(5):414.

Kisvarday, Z. F., Kim, D. S., Eysel, U. T., and Bonhoeffer, T. (1994). Relationship between lateral inhibitory connections and the topography of the orientation map in cat visual cortex. *Eur. J. Neurosci.*, 6:1619–1632.

Kisvarday, Z. F., Toth, E., Rausch, M., and Eysel, U. T. (1997). Orientation-specific relationship between populations of excitatory and inhibitory lateral connections in the visual cortex of the cat. *Cereb. Cortex*, 7:605–618.

Klier, K. (1972). Absorption and scattering in plane parallel turbid media. *J. Opt. Soc. Am.*, 62:882–885.

Klingler, J. and Fromherz, P. (1995). Profiles of voltage and of channel density in a planar membrane cable on micromachined silicon. *Ber. Bunsenges. Phys. Chem.*, 99:958–964.

Koch, C. and Segev, I. (1998). *Methods in neuronal modeling*. MIT Press, Cambridge, MS.

Koehler, B.-U. and Orglmeister, R. (1999). Independent component analysis using autoregressive models. In Cardoso, J.-F., Jutten, C., and Loubaton, P., editors, *Proceedings of the ICA99 workshop*, volume 1, pages 359–363.

Kreisman, N. R., LaManna, J. C., Liao, S.-C., Teh, E. R., and Alcala, J. R. (1995). Light transmittance as an index of cell volume in hippocampus slices: optical differences of interfaced and submerged positions. *Brain Res.*, 693:179–186.

Kuffler, S. W. (1953). Discharge patterns of and functional organization of mammalian retina. *J. Neurophysiol.*, 16:37–68.

Kussinger, M., Grünbauer, R., Stetter, M., and Lang, E. W. (1997). Prenatal development of feature maps with pure positive input activities. In Elsner, N. and Wässle, H., editors, *From membrane to mind.* Thieme Stuttgart.

Lee, T.-W. (1998). *Independent Component Analysis: Theory and Applications*. Kluwer Academic Publishers, Dordrecht.

Levitt, J. B. and Lund, J. S. (1997). Contrast dependence of contextual effects in primate visual cortex. *Nature*, 387:73–76.

Levitt, J. B., Lund, J. S., and Yoshioka, T. (1996). Anatomical substrates for early stages in cortical processing of visual information in the macaque monkey. *Behav. Brain Res.*, 76:5–19.

Li, Z. (1998). A neural model of contour integration in the primary visual cortex. *Neural Comput.*, 10:903–940.

Linden, D. J. and Connor, J. A. (1995). Long-term synaptic depression. *Annu. Rev. Neurosci.*, 18:319–357.

Linsker, R. (1989). How to generate ordered maps by maximizing the mutual information between input and output signals. *Neural Comput.*, 1:402–411.

Livingstone, M. S. and Hubel, D. H. (1984). Anatomy and physiology of a color system in the primate visual system. *J. Neurosci.*, 4:309–356.

Logothetis, N. (2000). Can current fMRI techniques reveal the micro-architecture of cortex? *Nature Neurosci.*, 3(5):413.

Logothetis, N. K., Pauls, J., Augath, M., Trinath, T., and Oeltermann, A. (2001). Neurophysiological investigation of the basis of the fMRI signal. *Nature*, 412:150–157.

Lund, J. S. (1987). Local circuit neurons of macaque monkey striate cortex: I. neurons of laminae 4c and 5a. *J. Comp. Neurol.*, 257:60–92.

Lund, J. S. (1988). Anatomical organization of macaque monkey striate visual cortex. *Annu. Rev. Neurosci.*, 11:253–288.

Lund, J. S., Levitt, J. B., and Wu, Q. (1994). Topography of excitatory and inhibitory connectional anatomy in monkey visual cortex. In Lawton, T. B., editor, *Computational vision based on neurobiology.*, pages 174–184.

Lund, J. S., Wu, Q., Hadingham, P. T., and Levitt, J. B. (1995). Cells and circuits contributing to functional properties in area v1 of macaque monkey cerebral cortex: bases for neuroanatomically realistic models. *J. Anat.*, 187:563–581.

Maass, W. (2000). Computation with spiking neurons. In Arbib, M. A., editor, *The Handbook of Brain Theory and Neural Networks*. MIT Press, Cambridge.

Maass, W. and Bishop, C. M. (1998). *Pulsed Neural Networks*. MIT Press Cambridge.

MacVicar, B. A. and Hochman, D. (1991). Imaging of synaptically evoked intrinsic optical signals in hippocampal slices. *J. Neurosci.*, 11:1458–1469.

Magistretti, P. J. and Pellerin, L. (1999). Cellular mechanisms of brain energy metabolism and their relevance to functional brain imaging. *Phil. Trans. R. Soc. Lond. B*, 354:1155–1163.

Makeig, S., Bell, A. J., Jung, T.-P., and Sejnowski, T. J. (1996). Independent component analysis of electroencephalographic data. In Touretzky, D., Mozer, M., and Hasselmo, M., editors, *Advances in Neural Information Processing Systems*, pages 145–151. MIT Press, Cambridge MA.

Malach, R., Amir, Y., Harel, M., and Grinvald, A. (1993). Relationship between intrinsic connections and functional architecture revealed by optical imaging and in vivo targeted biocytin injections in primate striate cortex. *Proc. Natl. Acad. Sci. USA*, 90:10469–10473.

Malach, R., Tootell, R. B., and Malonek, D. (1994). Relationship between orientation domains, cytochrome oxidase stripes, and intrinsic horntal conections in squirrel monkey area v2.nizo. *Cereb. Cortex*, 4:151–165.

Malonek, D. and Grinvald, A. (1996). Interactions between electrical activity and cortical microcirculation revealed by imaging spectroscopy: implications for functional brain mapping. *Science*, 272:551–554.

Marcos, S., Tornow, R. P., Elsner, A. E., and Navarro, R. (1997). Foveal cone spacing and cone photopigment density difference: objective measurement in the same subjects. *Vision Res.*, 37:1909–1915.

Markram, H., Lübke, J., Frotscher, M., and Sakman, B. (1997). Regulation of synaptic efficacy by coincidence of postsynaptic aps and epsps. *Science*, 275:231–215.

Markram, H. and Tsodyks, M. (1996). Redistribution of synaptic efficacy between neocortical pyramidal neurons. *Nature*, 382:807–810.

Markram, H., Wang, Y., and Tsodyks, M. (1998). Differential signalling via the same axon of neocortical pyramidal neurons. *Proc. Natl. Acad. Sci. USA*, 95:5323–5328.

Mayhew, J., Zheng, Y., Hou, Y., Vuksanovic, B., Berwick, J., Askew, S., and Coffey, P. (2000). Spectroscopic analysis of changes in remitted illumination: The response to increased neural activity in brain. *NeuroImage*, page in press.

Mayhew, J., Zheng, Y., Johnston, D., Coffey, P., Berwick, J., Porril, J., Stone, J., Stetter, M., Schießl, I., and Obermayer, K. (1999). Weak models in the analysis of optical imaging data: Haemoglobin changes following neural activity. *Soc. Neurosci. Abstr.*, 25:1638.

Mayhew, J. E. W., Askew, S., Zheng, Y., Porril, J., Westby, G. W. M., Redgraves, P., Rector, D. M., and Harper, R. M. (1996a). Cerebral vasomotion: 0.1hz oscillation in reflected light imaging of neural activity. *NeuroImage*, 4:183–193.

Mayhew, J. E. W., Hu, D., Zheng, Y., Askew, S., Hou, Y., Berwick, J., Coffey, P., and Brown, N. (1996b). An evaluation of linear model analysis techniques for processing images of micro-circulation activity. Technical Report 112, Artificial Intelligence Vision Research Unit, University of Sheffield, UK.

McCormick, D. A., Connors, B. W., Lighthall, J. E., and Prince, D. A. (1985). Comparative electrophysiology of pyramidal and sparsely spiny stellate neurons of the neocortex. *J. Neurophysiol.*, 54(4):782–806.

McKeown, M., Makeig, S., Brown, G., Jung, T., McKeown, S. J., Makeig, S., Brown, G. G., Jung, T. P., Kindermann, S. S., Bell, A., and Sejnowski, T. (1998). Analysis of fMRI by blind separation into independent spatial components. *Human Brain Mapping*, 6:160–188.

McLoughlin, N. P. and Blasdel, G. G. (1997). Effect of wavelength on differential images of ocular dominance and orientation in monkey striate cortex. *Soc. Neurosci. Abstr.*, 23:13.

McLoughlin, N. P. and Blasdel, G. G. (1998). Interactions between orientation, ocular dominance, and positional hypercolumns in monkey striate cortex. *Soc. Neurosci. Abstr.*, 24:9.

Mel, B. W. (1993). Synaptic integration in an excitable dendritic tree. *J. Neurophysiol.*, 70(3):1086–1101.

Miller, K. D. (1994). A model for the development of simple cell receptive fields and the ordered arrangement of orientation columns through activity-dependent competition between on- and off-center inputs. *J. Neurosci.*, 14:409–441.

Miller, K. D. and MacKay, D. J. C. (1994). The role of constraints in Hebbian learning. *Neural Comput.*, 6:100–126.

Molgedey, L. and Schuster, H. G. (1994). Separation of a mixture of independent signals using time delayed correlations. *Phys. Rev. Lett.*, 72:3634–3637.

Müller, T., Stetter, M., Hübener, M., Gödecke, I., Chapman, B., Löwel, S., Sengpiel, F., Bonhoeffer, T., and Obermayer, K. (2000). An analysis of orientation and ocular dominance patterns in the visual cortex of cats and ferrets. *Neural Comput.*, page in press.

Mundel, T., Dimitrov, A., and Cowan, J. D. (1996). Visual cortex circuitry and orientation tuning. In Mozer, M. C., Jordan, M. I., and Petsche, T., editors, *Neural Information Processing Systems*, volume 9. MIT Press Cambridge, MA.

Obermayer, K. and Blasdel, G. G. (1993). Geometry of orientation and ocular dominance columns in monkey striate cortex. *J. Neurosci.*, 13:4114–4129.

Obermayer, K. and Blasdel, G. G. (1997). Singularities in primate orientation maps. *Neural Comput.*, 9:555–567.

Ohzawa, I., Sclar, G., and Freeman, R. D. (1985). Contrast gain control in the cat's visual system. *J. Neurophysiol.*, 54:651–657.

Oja, E. (1997). The nonlinear pca learning rule in independent component analysis. *Neurocomputing*, 17:25–45.

O'Kusky, J. and Colonnier, M. (1982). A laminar analysis of the number of neurons, glia, and synapses in the visual cortex (area 17) of adult macaque monkeys. *J. Comp. Neurol.*, 210:278–290.

Olshausen, B. A. and Field, D. J. (1995a). Natural image statistics and efficient coding. In *Proceedings of the Workshop on Information Theory and the Brain, Sept. 4-5,*.

Olshausen, B. A. and Field, D. J. (1995b). Sparse coding of natural images produces localized, oriented, bandpass receptive fields. Technical report, Cornell University, Ithaca, New York.

Olshausen, B. A. and Field, D. J. (1996). Emergence of simple-cell receptive field properties by learning a sparse code for natural images. *Nature*, 381:607–609.

Ontrup, J. and Ritter, H. (1998). Perceptual grouping in a neural model: reproducing human texture perception. Technical Report SFB360-TR-98/6, Neuroinformatics Group, University of Bielefeld.

Orbach, H. S., Cohen, L. B., and Grinvald, A. (1985). Optical mapping of electrical activity in rat somatosensory and visual cortex. *J. Neurosci.*, 5:1886–1895.

Ostergaard, L., Smith, D. F., Vestergaard-Poulsen, P., Hansen, S. B., Gjedde, A. D., and Gyldensted, C. (1998). Absolute cerebral blood-flow and blood volume measured by magnetic resonance imaging bolus tracking: comparison with positron emission tomography. *J. Cereb. Blood Flow Metab.*, 18:425–432.

Otto, T., Stetter, M., Müller, T., Sengpiel, F., Hübener, M., Bonhoeffer, T., and Obermayer, K. (1998). Source separation of intrinsic signals from image sequences of cat area 17. In Elsner, N. and Wehner, R., editors, *New neuroethology on the move*, page 781. Thieme, Stuttgart 1998.

Papoulis, A. (1965). *Probability, random variables, and stochastic processes.* McGraw-Hill Book Company.

Paradiso, M. A. (1988). A theory for the use of visual orientation information which exploits the columnar structure of striate cortex. *Biol. Cybern.*, 58:35–49.

Pawelzik, K. R., Ernst, U., Wolf, F., and Geisel, T. (1996). Orientation contrast sensitivity from long-range interactions in visual cortex. In Mozer, M. C., Jordan, M. I., and Petsche, T., editors, *Neural Information Processing Systems*, volume 9. MIT Press Cambridge MA.

Pawley, J. B., editor (1995). *Handbook of Biological Confocal Microscopy.* Plenum Press, New York.

Piepenbrock, C., Weber, C., Obermayer, K., and Stetter, M. (1997). Topographic refinement and functional segregation mediated by nonlinear mechanisms of cortical development. *Soc. Neurosci. Abstr.*, 23:905.

Platt, J. C. and Faggin, F. (1992). Networks for the separation of sources that are superimposed and delayed. In Moody, J. E., Hanson, S. J., and Lippmann, R. P., editors, *Advances in Neural Information Processing Systems*, volume 4, pages 730–737.

Poenie, M. (1992). Measurement of intracellular calcium with fluorescent calcium indicators. *Neuromethods*, 20:129–174.

Polat, U., Mizobe, K., Pettet, M. W., Kasamatsu, T., and Norcia, A. M. (1998). Collinear stimuli regulate visual responses depending on cell's contrast threshold. *Nature*, 391:580–584.

Press, W. H., Flannery, B. P., Teukolsky, S. A., and Vetterling, W. T. (1988). *Numerical Recipes in C.* Cambridge University Press, Cambreidge, U.K.

Rall, W. (1964). Theoretic significance of dendritic trees for for neuronal input-output relations. In Reiss, R., editor, *Neuronal Theory and Modeling*, pages 73–97. Stanford University Press, Stanford CA.

Rall, W. (1967). Distinguishing theoretical synaptic potentials computed for different soma-dendritic distribution of synaptic inputs. *J. Neurophysiol.*, 30:1138–1168.

Rao, S. C., Toth, L. J., and Sur, M. (1997). Optically imaged maps of orientation preference in primary visual cortex of cats and ferrets. *J. Comp. Neurol.*, 387:358–370.

Rockland, K. S. and Lund, J. S. (1983). Intrinsic laminar lattice connections in primate visual cortex. *J. Comparative. Neurol.*, 216:303–318.

Rojer, A. S. and Schwartz, E. L. (1990). Cat and monkey cortical columnar pattern modeled by bandpass-filtered 2D white noise. *Biol. Cybern.*, 62:381–391.

Roland, P. E. (1993). *Brain activation.* John Wiley & Sons.

Ruderman, D. L. (1994). Designing receptive fields for highest fidelity. *Network*, 5:147–155.

Rüger, S. M. (1996). Stable dynamic parameter adaptation. In Touretzky, D. S., Mozer, M. C., and Hasselmo, M. E., editors, *Advances in Neural Information Processing Systems.*, volume 8, pages 225–231. MIT Press Cambridge, MA.

Sachse, S., Rappert, A., and Galizia, C. G. (1999). The spatial representation of chemical structures in the antennal lobe of honeybees: steps towards the olfactory code. *Eur. J. Neurosci.*, 11:3970–3982.

Salzberg, B. M., Davila, H. V., and Cohen, L. B. (1973). Optical recording of impulses in individual neurones of an invertebrate central nervous system. *Nature*, 246:508–509.

Schießl, I., Schöner, H., Stetter, M., Dima, A., and Obermayer, K. (2000a). Regularized second order source separation. In *Proceeding of the ICA*2000 workshop, Helsinki*, page in press.

Schießl, I., Stetter, M., Mayhew, J. E. W., Askew, S., McLoughlin, N., Levitt, J. B., Lund, J. S., and Obermayer, K. (1999). Blind separation of spatial signal patterns from optical imaging records. In Cardoso, J.-F., Jutten, C., and Loubaton, P., editors, *Proceedings of the 1. ICA99 Workshop, Aussois*, volume 1, pages 179–184.

Schießl, I., Stetter, M., Mayhew, J. E. W., McLoughlin, N., Lund, J. S., and Obermayer, K. (2000b). Blind signal separation from optical imaging recordings with extended spatial decorrelation. *IEEE Trans. Biomed. Engineering*, page in press.

Schießl, I., Stetter, M., Otto, T., Sengpiel, F., Hübener, M., Bonhoeffer, T., and Obermayer, K. (1998). Signal extraction from optical imaging data from cat area 17 by blind separation of sources. *Soc. Neurosci. Abstr.*, 24:9.

Schiller, P. H., Finlay, B. L., and Volman, S. F. (1976a). Quantitative studies of single-cell properties in monkey striate cortex. i. spatiotemporalorganization of receptive fields. *J. Neurophysiol.*, 39:1288–1319.

Schiller, P. H., Finlay, B. L., and Volman, S. F. (1976b). Quantitative studies of single-cell properties in monkey striate cortex. ii. orientation specificity and ocular dominance. *J. Neurophysiol.*, 39:1321–1333.

Schmidt, R. F. and Thews, G. (1989). *Human Physiology*. Springer, Berlin.

Schmitt, J. M. and Ben-Letaief, K. (1996). Efficient monte carlo simulation of confocal microscopy in biological tissue. *J. Opt. Soc. Am. A*, 13:952–961.

Scholz, M., Stetter, M., and Obermayer, K. (1998). Deblurring of confocal microscope images: tissue properties, optics and simulations. *Soc. Neurosci. Abstr.*, 24:1060.

Schöner, H., Stetter, M., Schießl, I., Mayhew, J. E. W., Lund, J. S., McLoughlin, N., and Obermayer, K. (1999a). Blind separation of noisy mixtures by iterative decorrelation.. Learning workshop, Snowbird, Utah, April 6-9, 1999.

Schöner, H., Stetter, M., Schießl, I., Mayhew, J. E. W., Lund, J. S., McLoughlin, N., and Obermayer, K. (1999b). Noise-robust blind separation of sources for optiocal imaging of intrinsic signals. *Soc. Neurosci. Abstr.*, 25:783.

Schöner, H., Stetter, M., Schießl, I., Mayhew, J. E. W., Lund, J. S., McLoughlin, N., and Obermayer, K. (2000). Application of blind separation of sources to optical recording of brain activity. In Leen, T. K. and Bottou, L., editors, *Neural Information Processing Systems*, page in press.

Schreiber, W. F. (1956). The measurement of third order probability distributions of television signals. *IRE Trans. Inform. Theory*, IT-2:94–105.

Schuman, E. M. and Madison, D. V. (1994). Nitric oxide and synaptic function. *Annu. Rev. Neurosci.*, 17:153–183.

Schutter, E. D. (1999). Using realistic models to study synaptic integration in cerebellar Purkinje cells. *Rev. Neurosci.*, 10:233–245.

Schutter, E. D. and Bower, J. M. (1993). Sensitivity of synaptic plasticity to the $ca^2+$ permeability of nmda channels: A model of long-term potentiation in hippocampal neurons. *Neural Computation*, 5:681–694.

Schwabe, L., Stetter, M., and Obermayer, K. (2000). Reduced spike-frequency adaptation in attentive states favoures the representation of salient features in a computational model of the primary visual cortex. *Soc. Neurosci. Abstr.*, 26:142.

Sclar, G. and Freeman, R. D. (1982). Orientation selectivity in the cat's striate cortex is invariant with stimulus contrast. *Exp. Brain Res.*, 46:457–461.

Segev, I. (1995). Dendritic processing. In Arbib, M. A., editor, *The handbook of brain theory and neural networks*. MIT Press, Cambridge, MA.

Segev, I. and Parnas, I. (1983). Synaptic integration mechanisms: a theoretical and experimental investigation of temporal postsynaptic interactions between excitatory and inhibitory inputs. *Biophys. J.*, 41:41–50.

Sengpiel, F., Baddeley, R. J., Freeman, T. C., Harrad, R., and Blakemore, C. (1998a). Different mechanisms underlie three inhibitory phenomena in cat area 17. *Vision Res.*, 38:2067–2080.

Sengpiel, F., Gödecke, I., Strawinski, P., Hübener, M., Löwel, S., and Bonhoeffer, T. (1998b). Innate and environmental factors in the formation of functional maps in cat visual cortex. *Neuropharmacology*, 37:607–621.

Sengpiel, F., Sen, A., and Blakemore, C. (1997). Characteristics of surround inhibition in cat area 17. *Exp. Brain Res.*, 116:216–228.

Shannon, C. E. (1948). A mathematical theory of communication. *Bell Syst. Tech.*, 27:379.

Shepherd, G. M. (1988). *Neurobiology*. Oxford University Press, Oxford.

Shmuel, A. and Grinvald, A. (1996). Functional organization for direction of motion and its relationship to orientation maps in cat area 18. *J. Neurosci.*, 16:6945–6964.

Sillito, A. M., Grieve, K. L., Jones, H. E., Cudeiro, J., and Davis, J. (1995). Visual cortical mechanisms detecting focal discontinuities. *Nature*, 378:492–496.

Skottun, B. C., Bradley, A., Sclar, G., Ohzawa, I., and Freeman, R. D. (1987). The effects of contrast on visual orientation and spatial frequency discrimination: a comparison of single cells and behaviour. *J. Neurophysiol.*, 57:773–786.

Sliney, D. and Wolbarsht, M. (1980). Safety with lasers and other optical sources. Plenum Press, New York.

Somers, D. C., Nelson, S. B., and Sur, M. (1995). An emergent model of orientation selectivity in cat visual cortical simple cells. *J. Neurosci.*, 15:5448–5465.

Steriade, M., McCormick, D. A., and Sejnowski, T. J. (1993). Thalamocortical oscillations in the sleeping and aroused brain. *Science*, 262:679–685.

Stetter, M. (1994). *Selbstorganisation funktioneller Strukturen im primären visuellen Kortex.* Roderer, Regensburg.

Stetter, M., Adorján, P., Bartsch, H., and Obermayer, K. (1998a). Modelling contrast-adaptation and contextual effects in primary visual cortex. In *Proceedings of the ICONIP '98 conference, Kitakyushu*, volume 2, pages 669–672.

Stetter, M., Bartsch, H., and Obermayer, K. (2000a). A mean field model for orientation tuning, contrast saturation and contextual effects in the primary visual cortex. *Biol. Cybern.*, 82:291–304.

Stetter, M., Greve, H., Galizia, C. G., and Obermayer, K. (2000b). Analysis of calcium imaging signals from the honeybee brain by nonlinear models. *NeuroImage*, page submitted.

Stetter, M., Kussinger, M., Schels, A., Seeger, E., and Lang, E. W. (1995a). Self-organization of cortical receptive fields and columnar structures in a Hebb-trained neural network. In Mira, J. and Sandoval, F., editors, *From natural to artificial neural computation.*, pages 37–44. Springer Berlin.

Stetter, M. and Lang, E. W. (1994). A model for the prenatal formation of simple orientation selective cells. *Invest. Ophthalm. Vis. Sci.*, 35(4):1774.

Stetter, M., Lang, E. W., and Müller, A. (1993). Emergence of orientation selective simple cells simulated in deterministic and stochastic neural networks. *Biol. Cybern.*, 68:465–476.

Stetter, M., Lang, E. W., and Obermayer, K. (1997a). Synapse clustering can drive simultaneous ON-OFF and ocular dominance segregation in a model of area 17. In Gerstner, W., Germond, A., Hasler, M., and Nicoud, J.-D., editors, *Artificial Neural Networks - ICANN '97*, volume 1327 of *Lecture notes in computer science*, pages 189–194. Springer Berlin.

Stetter, M., Lang, E. W., and Obermayer, K. (1998b). Unspecific long-term potentiation can evoke functional segregation in a model of area 17. *NeuroReport*, 9:2697–2702.

Stetter, M., Lang, E. W., and Obermayer, K. (1998c). Unspecific LTP may lead to simultaneous and robust ON-OFF and eye-dominance segregation correlation-based learning models. Technical report, Machines that learn workshop, Snowbird, Utah, USA.

Stetter, M., Lang, E. W., Weih, P., and Obermayer, K. (1998d). Simultaneous stabilization of ON-OFF and ocular dominance patches by modeled unspecific LTP. *Eur. J. Neurosci*, 10(S10):332.

Stetter, M., Müller, A., and Lang, E. W. (1994). Neural network model for the coordinated formation of orientation preference and orientation selectivity maps. *Phys. Rev. E*, 50:4167–4181.

Stetter, M. and Obermayer, K. (1998). Tissue optics simulations of scanning laser techniques for optical imaging. In Elsner, N. and Wehner, R., editors, *New Neuroethology on the Move*, page 779. Thieme, Stuttgart.

Stetter, M. and Obermayer, K. (1999). Simulation of scanning laser techniques for optical imaging of blood-related intrinsic signals. *J. Opt. Soc. Am. A*, 16:58–70.

Stetter, M., Obermayer, K., and Lang, E. W. (1997b). A model for simultaneous ON-OFF and ocular dominance segregation in area 17 via synaptic clustering. In Elsner, N. and Wässle, H., editors, *From Membrane to Mind*. Thieme Stuttgart.

Stetter, M., Otto, T., Müller, T., Sengpiel, F., Hübener, M., Bonhoeffer, T., and Obermayer, K. (1997c). Temporal and spatial analysis of intrinsic signals from cat visual cortex. *Soc. Neurosci. Abstr.*, 23:455.

Stetter, M., Schels, A., and Lang, E. W. (1995b). A neural network model for the prenatal formation of orientation preference and -selctivity maps. *Invest. Ophthalm. Vis. Sci.*, 36(4):911.

Stetter, M., Schels, A., and Lang, E. W. (1995c). Prenatal self-organization of orientation maps simulated in a neural network. In Elsner, N. and Menzel, R., editors, *Learning and memory*. Thieme Stuttgart.

Stetter, M., Schießl, I., Otto, T., Sengpiel, F., Hübener, M., Bonhoeffer, T., and Obermayer, K. (2000c). Principal component analysis and blind separation of sources for optical imaging of intrinsic signals. *NeuroImage*, 11:482–490.

Stetter, M., Sendtner, R. A., Fitzke, F. W., and Gabel, V. P. (1995d). Development of a kinetic SLO microperimetry module for functional macular diagnostics. *Vision Res.*, 35:S43.

Stetter, M., Sendtner, R. A., and Timberlake, G. T. (1996). A novel method for measuring saccade profiles using the Scanning Laser Ophthalmoscope. *Vision Res.*, 36:1987–1994.

Stetter, M., Timberlake, G. T., Sendtner, R. A., and Webb, R. H. (1995e). SLO saccade profile measurements and the effects of retinal raster size and distortion. In *SPIE, Proceedings of the European Biomedical Optics Symposium*, volume 2623, pages 98–109.

Stetter, M., Weber, C., and Obermayer, K. (1998e). Depth-resolved scanning laser optical imaging – a Monte Carlo simulation study. *Soc. Neurosci. Abstr.*, 24:1137.

Timberlake, G. T. and Stetter, M. (1996). Magnification and distortion in the Scanning Laser Ophthalmoscope. *Invet. Ophthalm. Vis. Sci.*, 37(4):100.

Todorov, E., Siapas, A., and Somers, D. (1996). A model of recurrent interactions in primary visual cortex. In Mozer, M. C., Jordan, M. I., and Petsche, T., editors, *Advances in Neural Information Processing Systems*, volume 8. MIT Press Cambridge, Massachusetts.

Tootell, R. B., Silverman, M. S., Hamilton, S. L., Switkes, E., and DeValois, R. L. (1988). Functional anatomy of macaque striate cortex. V. spatial frequency. *J. Neurosci.*, 8:1610–1624.

Tornow, R.-P., Beuel, S., and Zrenner, E. (1997). Modifying a Rodenstock Scanning Laser Ophthalmoscope for imaging densitometry. *Applied Optics*, 36:5621–5629.

Tsien, R. W. and Tsien, R. Y. (1990). Calcium channels, stores, and oscillations. *Annu. Rev. Cell Biol.*, 6:715–760.

Tsien, R. Y. (1980). New calcium indicators and buffers with high selectivity against magnesium and protons: design, synthesis, and properties of prototype structures. *Biochemistry*, 19:2396.

Ts'o, D. Y., Frostig, R. D., Lieke, E. E., and Grinvald, A. (1990). Functional organization of primate visual cortex revealed by high resolution optical imaging. *Science*, 249:417–420.

Ts'o, D. Y. and Gilbert, C. D. (1988). The organization of chromatic and spatial interactions in the primate striate cortex. *J. Neurosci.*, 8:1712–1727.

Tsodyks, M. V. and Sejnowski, T. (1995). Rapid state switching in balanced cortical network models. *Network*, 6:111–124.

Van Essen, D. C., Anderson, C. H., and Felleman, D. J. (1992). Information processing in the primate visual system: An integrated systems perspective. *Science*, 255:419–423.

van Gemert, M. J. C., Jacques, S. L., Sterenborg, H. J. C. M., and Star, W. M. (1989). Skin optics. *IEEE Trans. Biomed. Eng.*, 36:1146–1154.

Verkhratsky, A. and Kettenmann, K. (1996). Calcium signalling in glial cells. *Trends Neurosci.*, 19:346–352.

Vetter, R., Vesin, J., Celka, P., and Scherrer, U. (1999). Observer of the autonomic cardiac outflow in humans using non-causal blind source separation. In Cardoso, J.-F., Jutten, C., and Loubaton, P., editors, *Proceedings of the 1. ICA99-Workshop, Aussois*, pages 161–166.

Volgushev, M., Pernberg, J., and Eysel, U. T. (2000). Comparison of the selectivity of postsynaptic potentials and spike responses in cat visual cortex. *Eur. J. Neurosci.*, 12:257–263.

Vollgraf, R., Stetter, M., and Obermayer, K. (2000). Convolutive decorrelation procedures for blind source separation. In *Proceeding of the ICA*2000 workshop, Helsinki*, page in press.

Wässle, H., Grünert, U., Röhrenbeck, J., and Boycott, B. B. (1989). Cortical magnification factor and the ganglion cell density of the primate retina. *Nature*, 341:643–646.

Webb, R. H. and Hughes, G. W. (1981). Scanning laser ophthalmoscope. *IEEE Transactions on Biomedical Engineering*, 28:488–492.

Weliky, M., Bosking, W. H., and Fitzpatrick, D. (1996). A systematic map of direction preference in primary visual cortex. *Nature*, 379:725–728.

Wersing, H., Steil, J., and Ritter, H. (1997). A layered recurrent neural network for feature grouping.. In Gerstner, W., Germond, A., Hasler, M., and Nicoud, J.-D., editors, *Artificial Neural Networks - ICANN '97*, volume 1327 of *Lecture notes in computer science*, pages 439–444. Springer Berlin.

Wersing, H., Steil, J. J., and Ritter, H. (2000). A competitive layer model for feature binding and sensory segmentation. *Neural Comput.*, page in press.

Williamson, A. M., Ohara, P. T., and Ralston, H. J. (1993). Electron microscopic evidence that cortical terminals make direct contact onto cells of the thalamic reticular nucleus in the monkey. *Brain Res.*, 631:175–179.

Wilson, B. C. and Jacques, S. L. (1990). Optical reflectance and transmittance of tissues: Principles and applications. *IEEE J. Quant. Electron.*, 26:2186–2199.

Wiser, A. K. and Callaway, E. M. (1996). Contributions of individual layer 6 pyramidal neurons to local circuitry in macaque primary visual cortex. *J. Neurosci.*, 16:2724–2739.

Yabuta, N. H. and Callaway, E. M. (1998). Functional streams and local connections of layer 4c neurons in primary visual cortex of the macaque monkey. *J. Neurosci.*, 15:9489–9499.

Yoshioka, T., Blasdel, G. G., Levitt, J. B., and Lund, J. S. (1996). Relation between patterns of intrinsic lateral connectivity, ocular dominance, and cytochrome oxidase-reactive regions in macaque monkey striate cortex. *Cerebral Cortex*, 6:297–310.

Yoshioka, T., Levitt, J. B., and Lund, J. S. (1994). Independence and merger of thalamocortical channels within macaque monkey primary visual cortex: Anatomy of interlaminar projections. *Vis. Neurosci.*, pages 467–489.

Yuille, A. L., Kammen, D. M., and Cohen, D. S. (1989). Quadrature and the development of orientation selective cortical cells by Hebb-rules. *Biol. Cybern.*, 61Scofield CL:183–194.

Ziegaus, C. and Lang, E. W. (1999). Neural implementation of the JADE-algorithm. In *Lecture Notes in COmputer Science*, volume 1607, pages 487–496.

Zohary, E. (1992). Population coding of visual stimuli by cortical neurons tuned to more than one dimension. *Biol. Cybern.*, 66:265–272.

# Index